P9-AGP-543

DATE DUE

NO 27 '95			
DE 20 '00			
JE 7 '03			

The Sick
Building
Syndrome

The Sick Building Syndrome

How Indoor Air Pollution
Is Poisoning Your Life—
And What You Can Do

Nicholas Tate

NEW
HORIZON
PRESS

Far Hills,
New Jersey

Requests for permission should be addressed to:
New Horizon Press
P.O. Box 669
Far Hills, NJ 07931

Tate, Nicholas.
    The sick building syndrome

Library of Congress Catalog Card Number: 93-84519

ISBN: 0-88282-085-0 (hc)
ISBN: 0-88282-082-6 (pb)
New Horizon Press

Manufactured in the U.S.A.

1998 1997 1996 1995 1994 / 5 4 3 2 1

CONTENTS

ACKNOWLEDGMENTS

It's been said that success has a hundred fathers, while failure is an orphan. To the extent that the completion of this book marks a success of sorts, the author gratefully acknowledges the generous assistance of many individuals whose input and support have helped see this project to its completion.

This book could not have been written without the time and resources provided by the Harvard School of Public Health through its Journalism Fellowship for Advanced Studies in Public Health. Much of the work was conducted while I was a Visiting Fellow in 1991 and 1992. I am deeply indebted to fellowship director, Robert Meyers, whose keen insight and guidance were critical in the early, middle, and final stages of the manuscript. Many others within the Harvard community provided invaluable assistance in this document's preparation. My primary mentor and advisor was John Spengler, who went above and beyond the call of duty in generously offering his time, research findings, and expert opinions as both a guide and technical reviewer. Others at Harvard who made significant contributions to the book's completion included Joseph Brain, John Graham, P. Barry Ryan, Dean Harvey Fineberg, and Assistant Dean Jay Winsten.

I have also benefited from the suggestions of many IAQ specialists and environmental health experts who read parts of the manuscript, opened their files to me, and made other contributions, in particular Nicholas Ashford, Jeffrey May, John McCarthy, Paul Marcus, Joseph P. Kennedy II, Mary Beth Smuts, Paul Keough, Robert Axelrad, Marilyn Black, Gerald Ross, Lance Wallace, Stan Salisbury, Max Kiefer, and others at

Acknowledgments

the EPA, NIOSH, OSHA, the GAO, and the American Lung Association. Special thanks also to Joan Dunphy at New Horizon for her care and interest in this project; to my agent, Richard Curtis; and to the thousands of researchers whose studies, articles, books, and other writings informed the chapters of this book.

In addition, the efforts of several editors and colleagues at the *Boston Herald* cannot go unacknowledged, especially those of Alan Eisner, Andrew Costello, James MacLaughlin, Andrew Gully, and Kevin Convey. And a number of other individuals made significant contributions: Nicholas and Annette Tate, Victor and Miriam Newman, Rebecca and Brian Dowsey, Christopher Tate, Rob and Joan Newman, Frank and Rebecca Faiola, the Mandells, the Hills, the Jacobys, the Schnurs, and the Marianos.

Last, but certainly not least, I want to express my deep gratitude to my wife, Jodi, who probably deserves coauthor status. During the demanding three years of this project's development—a period that included the birth of our son, Daniel—she provided unwavering support and encouragement.

INTRODUCTION

Air pollution.

We tend to associate the phrase with the outdoor environment—with factory smokestacks, automobile exhaust, urban smog.

But scientists and health researchers are beginning to advise that we rethink that long-held association as an emerging body of evidence increasingly suggests the nation's most pressing air pollution problems are actually found indoors—within the confines of our own homes, offices, schools, and workplaces.

New government studies show that the air inside the typical home harbors a witch's brew of dangerous air pollutants at levels that are many times higher than those found outdoors. Naturally occurring radon gas, present at unsafe levels in one out of every fifteen homes, will cause 14,000 lung-cancer deaths this year alone in America, according to United States Environmental Protection Agency estimates. Secondhand smoke will kill another 3,000 adults and send 11,000 children to the hospital with respiratory ailments. In Europe and the Far East, where smokers are even more prevalent, much the same situation occurs. Even the air in tobacco-free homes with low radon levels can be laced with up to 150 contaminants—from stove gases, furnaces, solvents, paints, furnishings, mold, and pesticides.

Unfortunately, the picture doesn't improve much when it comes to the workplace. Officials estimate up to 30 percent of all new buildings display classic "sick building syndrome" symptoms. Prevalence studies have suggested that ten to twenty-five million office workers—that's one in five—may suffer health problems caused by indoor air pollution sources and inadequate building ventilation. And the economic costs to the nation, from health care expenditures and lost worker productivity due to poor indoor air quality, are believed to total tens of billions of dollars each year.

In the face of these and other troubling findings, the EPA has consistently ranked indoor air pollution among the nation's top five environmental health problems, with the risks deemed far greater than those posed by toxic waste dumps, industrial emissions, and agricultural pesticide use. Yet indoor air programs continue to languish near the bottom of the federal government's funding and regulatory priorities, garnering far less attention than such high-visibility environmental issues as hazardous waste cleanup, wetlands protection, outdoor air quality, and water pollution. Consequently, health researchers' findings on indoor radon, secondhand smoke, pesticides, volatile organic compounds, and other pollutants have only recently begun to reach beyond the limited scope of academic labs, government reports, and obscure technical journals.

Not all the news is bad, however.

Growing concerns over the ailing health care system are increasingly shifting the focus of public health debates to *cause* and *prevention* of disease—a shift in which indoor air quality and other environmental factors are certain to figure prominently over the next decade. EPA funding for indoor air programs is steadily, if slowly, rising each year—from just $350,000 in 1987 to $7 million in 1993—as the agency moves to align its funding priorities with risk-assessment projections.

In the United States, Congress has begun to turn its attention to the issue, with comprehensive indoor air bills now pending in the House and Senate. And even business and labor leaders—notably those in building trades, real estate, and product manufacturing industries—are taking voluntary steps to modify industrial processes and practices to reduce health risks, as federal regulators grapple with competing proposals for addressing indoor air quality.

All of these factors suggest health researchers' lingering concerns about indoor air pollution are finally beginning to dawn on policy makers and corporate leaders. And, if current trends continue, it appears likely the nation will have some kind of federal indoor air quality program in place by the end of the decade. Yet as the public policy debate over the issue continues to rage, precious little information is filtering down to the average consumer, homeowner, and office worker about the risks of indoor air pollution and the practical steps that can be taken to limit exposure indoors.

To a large extent, this book has been written to fill that information gap and extend the reach of the public and private researchers, scientists, government officials, industry leaders, and health experts who are trying to get to the bottom of the problems facing the indoor environment.

The intent is to provide some useful advice on maintaining clean air quality—based on the most up-to-date research—and summarize the most important findings in the field in a readable, non-alarmist manner that helps place the problem in the context of other pressing environmental concerns.

The goal is to bridge the gap between science and the public, as policy makers evaluate the many technical, political, ethical, economical, and public health issues surrounding sick building syndrome and indoor air pollution.

The hope is that the information in this book will spark greater public interest in the issue and be put to practical use in the homes, offices, and other buildings where the work of scientific researchers can have the most profound impact.

CHAPTER 1

The End of a Dream, the Beginning of a Nightmare

When Sue Pitman and her husband, Lockett, bought their first new home in 1977, it was the completion of a dream.

Designed and built to their own specifications, the house seemed the ideal place for the couple to raise their two young children, Kyle and Gwin. Every detail, from the hand-picked kitchen cabinets to the plush carpeting and wallpaper, reflected the Pitmans' personal tastes and lifestyle. Situated in a quiet family neighborhood overlooking a postcard-perfect lake, near a large city, the house seemed almost too good to be true.

"It was like this little fantasy," she recalls. "This was the first time we'd actually lived in a brand-new house. It was built right on the lake. And because we built the house, we were doing all the decorating ourselves and got to choose a lot of the finishing touches and things inside."

1

Unfortunately, trouble was brewing in the Pitmans' suburban paradise even before they moved in. The first signs of it appeared midway through the family's first winter there, when Sue began experiencing chronic blinding headaches unlike anything she'd ever known. The headaches were followed by a string of respiratory illnesses, including asthma and pneumonia, also new to her. Then the Pitman children started getting sick. By the end of that first winter, all three had fallen into a troubling pattern of chronic colds, ear infections, headaches, and sore throats. "We were sick all the time," Sue says. "I was constantly taking the kids to doctors, and they'd give us medication, but nothing seemed to work."

Curiously, Lockett Pitman—a workaholic who, as an executive for a large computer firm, logged long hours outside the home—seemed to be immune to the health problems plaguing his wife and kids. Although that fact seemed strange to the Pitmans at the time, it was an obvious first clue to the cause of the family's ailments.

Personal Investigations

The next two years in their home were a blur. By their second winter, Kyle, then two years old, was diagnosed with asthma and a multitude of food allergies. Meanwhile, six-year-old Gwin had developed a variety of health problems, including allergies to dust and mold, and was experiencing frequent colds, stomach aches, and sore throats. Sue was suffering with her own set of bizarre afflictions, including frequent bladder and vaginal infections, bloody noses, rashes, nausea, allergies, and colds.

Doctors—and there were many over the family's first three years in the house—couldn't explain the multitude of symptoms. Pills and medication physicians prescribed were no help. "We saw a lot of doctors and pediatricians," Sue says. "Some of them thought it might be severe allergies or immune system things, but the tests always came up with nothing. They'd just keep giving us antibiotics, and they did nothing to help."

Frustrated, Sue eventually turned to the local library and started reading up on allergies, asthma, and immune system disorders. Several books recommended exercise and dietary changes, so she put the family on a workout regimen and switched to homegrown and organic fruits and vegetables.

Still, the problems persisted.

In the fall of 1980, Sue stumbled across a newspaper article on a new and controversial branch of medicine that centered on the links between health and environmental factors. The discipline, known as environmental medicine, was pioneered by one Dr. Theron Randolph. Feeling she had nothing to lose, she called Randolph's office and set up an appointment.

It was a decision that would profoundly change her life.

Over a period of months, Randolph performed a battery of tests that revealed Sue Pitman to be "chemically sensitive" as a result of near-constant exposure to toxic chemicals and pesticides in her immediate environment. Under Randolph's testing, she displayed allergy-like reactions to everything from natural gas, car exhaust fumes, and environmental tobacco smoke to alcohol, formaldehyde, dust, mold, and chemical pesticides. She also tested positive for simple allergies to corn, wheat, eggs, and milk. In time, she brought both her children to Randolph for testing, with similar results.

At the time, Sue says it was difficult for her to believe what the doctor was telling her. Harder still was accepting Randolph's harsh conclusion about what was causing the family's problems: the Pitmans' spanking-new dream home. But the evidence was striking. Among the first things Randolph determined was that Pitman and her kids were "reactive" to formaldehyde, a ubiquitous chemical preservative used in many building materials, adhesives, furnishings, and particleboard. Some experts believe formaldehyde is particularly noxious because it may have the potential to "sensitize" persons—meaning it might be one of a handful of chemicals that can act as precursors to "chemical sensitivities." Not surprisingly, the Pitmans' new home was virtually awash in the stuff, notably from the building materials, finishings, and the spiffy new kitchen and bathroom cabinets they had so carefully chosen. "I had really wanted those white formica-clad cabinets in the kitchen and the bathrooms," Sue recalls. "But I didn't know that they had a lot of formaldehyde in them. So I suppose they had something to do with it."

Formaldehyde was by no means the family's only problem, however. Like millions of other homes built over the past twenty years, the Pitman house was constructed with many synthetic, chemically treated building materials and finishings instead of more stable non-toxic alternatives such as wood, steel, and stone. Consequently, toxic compounds from the synthetic carpeting, vinyl wallpaper, chemically treated materials, and furnishings were giving off gasses and fouling the indoor environment. Moreover, because the house was a model of energy efficiency, featuring tight door frames, double-paned windows, and few "settling" cracks, there was little ventilation in the home. As a result, contaminants were being trapped indoors.

To make matters worse, the Pitmans' home was nestled in a neighborhood that was heavily treated with pesticides and chemicals to control insects and weeds. The pastoral lake out back was laden with contaminants from frequent applications of Malathion and other pesticides by local mosquito control officials. Similarly, routine chemical lawn-care treatments were at the root of the emerald lawns of the homes ringing the lake. As a result, the Pitmans couldn't simply open the doors and windows of their house to ventilate the home and clear the indoor air of chemicals without risking an influx of pesticides from the outdoor air.

A Bitter Pill

To address the Pitmans' problem, Randolph prescribed a bitter pill. He advised them to sell their tightly sealed new home and move to an older house in a more pristine neighborhood. Short of that, he recommended creating an air quality "oasis" in their home for sleeping, free of dust, mold and chemically treated items and furnished only with cotton beds, bedding, and air filters. He also placed the Pitmans on a special diet of natural foods and urged them to rid their home of any cleaning products, personal care items, and furnishings made with plastics, chemicals, or other synthetics. In 1981, the Pitmans tried to implement every one of Randolph's recommendations short of selling their home. They bought air filters for every room, created a sleeping "oasis," and improved their diet. They even convinced the local lawn care companies and mosquito-spraying agencies to cut back on their use of pesticides and provide advanced notice of their activities. The problems persisted.

Finally, in 1982, the family put their home up for sale and moved to an older and more drafty house across town. There they attempted to strip away as many of the Twentieth-Century vestiges as they could and live as simply as possible. They kept their home free of chemically treated furniture and other products, replaced their gas appliances with electric ones, sealed off the attached garage from the rest of the house, and created an "oasis" for sleeping. For a while, things seemed to improve, but the family was never completely healed. Sue Pitman has vivid memories of her husband bringing home particular items—newspapers or books, for instance—that would trigger reactions that sent her to the hospital emergency room. She also recalls nights that the air was so bad, she and her kids would sleep in the home's large tiled bathroom, cleared of all items but cotton bedding and loaded with air filters.

The family lived this way for two years. But episodes of asthma attacks and reactions lingered, often as a result of exposures away from home. Eventually, the Pitmans came to the realization that they would have to find somewhere else to live—another community, relatively free from the pesticides and other chemicals they encountered at every turn in the suburbs.

"We realized it just wasn't going to work there, no matter what we did," she says. "Even though we were doing much better, we were still limping along and vulnerable to what else was going on around us. We just couldn't get away from the chemicals."

"The kids would go to a birthday party and come home with problems, and we'd call and find out that the family had just had the house treated with pesticides. Or they'd go to school and come home with problems because the school yard had just been sprayed. I didn't want my children to be like that all their lives. I was determined to try to find a cure for my children."

Again, Sue turned to books, looking for places in the West and Southwest that might have cleaner air. She also did some networking with members of a support group for chemically sensitive persons she met through Randolph's office. Her research led her to Wimberley, Texas, a tiny desert town outside Austin where a small enclave of chemically sensitive people— including a friend of Sue's—had taken up residence and formed a support community.

In the spring of 1984, the Pitmans made the trek to Wimberley to look the place over. While there she discovered the chief reason Wimberley had sprung up as a center for chemically sensitive people in the late 1970s: the Environmental Health Center of Dallas, one of the nation's best known environmental medicine clinics, was a five-hour drive away. Sue was also pleased to discover a large natural food store had opened near the town's center and the air in the rugged Texas hill country of Wimberley was dry, clear, and desert-pure. That summer, Pitman and her children rented a Wimberley house. By the fall of 1984, the change in their health was so great that the Pitmans said goodbye to the suburbs for good and began building a new home in Texas.

Desert Detox

Today, the Pitmans are among some thirty chemically sensitive people living in the Wimberley community enclave. Their lifestyle bears little resemblance to the one they enjoyed in their old home a decade ago. Even their home, in the middle of ten acres of land, is a world away from their dreamhouse.

Built primarily of untreated Colorado Spruce logs, the house rises on stilts ten feet off the ground, allowing them to keep bugs and termites away without the use of pesticides. From the outside, it looks like an old prairie house straight out of the Old West. Inside, the picture is much the same. As in the homes of early pioneers, the Pitmans' dwelling is divided into two principal regions—the "safe area," used as sleeping quarters, which is furnished only with non-toxic items made from wood, steel, stone, or undyed cotton fabrics; and the "toxic area," which houses the kitchen, modern appliances, books, magazines, televisions, stereo equipment, and other common household items.

The two sections, joined by a breezeway, allow the family to "do normal things," Pitman says, "but we also have a safe area, or oasis, in the other part of the house, so you can get away from those things if you need to." The house is also outfitted with windows on every side and features an enormous screened-in porch, which the Pitmans sometimes use as sleeping quarters. Because the Texas climate is so mild, the family can sleep in the open air all but six weeks out of the year.

Soon after moving to Wimberley, Sue Pitman and her children began to feel an almost immediate improvement in their health. The change has been so dramatic, in fact, that the family is now able to tolerate increasing levels of chemical exposures. Kyle and Gwin, who were initially schooled at home to avoid pesticides and chemicals in classroom supplies in the Wimberley schools, are now able to attend classes like other children.

"We've all gotten so much better that we have started getting back into life," says Sue, now forty-eight years old,

7

crediting the family's non-toxic diet, air, and lifestyle with the turnaround. "We've gotten much stronger over time. For several years there, Kyle would spend days in bed, unable to move or read, after getting exposed to pesticides at school or when they'd spray the highway. But today he doesn't have those same reactions. He still reacts somewhat, but the reactions are much less severe."

The Pitmans have clearly paid a price for their newfound freedom. Lockett Pitman traded in his lucrative career as a computer executive for a home on the range. The family left many friends and relatives behind. And the Pitmans are rarely able to engage in simple family activities such as going out to the movies or for a pizza. But Sue Pitman says the sacrifices have been worth the change in her children's health.

"It was the best move we ever made," she asserts. "What we've been able to do is have a lifestyle that is just fine, where we walk a fine line between exposures and normal life. Our life is changed because it's just not worth feeling badly to do the things most people do or go into some of these places most people go."

"I feel like I'm 90 percent better most days. And I can't ever see a reason that I'd want to be in the situation we were in again. A while back we were looking for a small house in Austin so during the week Kyle can stay there while going to school, but I felt sick just looking inside the houses. My fantasy is to one day find a place in town that I can clean up and filter and use, at least some of the time. But I'm quickly discouraged when I start looking. I feel like I've just found a path of living that works for me and I reach out to participate in the world where I can."

An Emerging Problem

The experiences of Sue Pitman and her children are certainly unusual. But practitioners of environmental medicine say they bear a striking resemblance to those of millions of other

Americans who suffer from multiple chemical sensitivity, or MCS. A handful of pioneering researchers have been investigating the phenomenon for decades, with Randolph penning articles about allergic-type reactions to solvents, pesticides, and chemicals among patients as far back as the 1950s.[1] But the emerging condition, sometimes called "environmental illness," remains highly controversial and is still at the center of an intense and bitter medical debate.

The American Medical Association and many other mainstream medical organizations both in America and in other countries have refused to acknowledge that MCS is a bona fide illness. But environmental medicine practitioners argue that MCS is not only real but on the rise in industrialized nations as a result of increasingly polluted outdoor and indoor environments.

"We are seeing more and more patients who are reporting adverse responses to very low levels of substances we are exposed to all the time," says Dr. Gerald Ross, an environmental medicine specialist at the Environmental Health Center of Dallas, who believes as many as one in five Americans may suffer from MCS.

"There are a lot of people out there who may have some degree of MCS but don't recognize it, but no one has done any good prevalence studies of this," he adds. "The National Academy of Sciences has called for one, the EPA has called for it, and so has the Canadian government. But nobody's done it."

A meaningful definition of MCS has also eluded researchers to date. The most comprehensive comes from the American Academy of Environmental Medicine, which represents about half of the nation's one thousand-plus environmental medicine practitioners: "Ecologic illness is a poly-symptomatic, multi-system chronic disorder manifested by adverse reactions to environmental excitants, as they are modified by individual susceptibility in terms of specific adaptations. The excitants are present in air, water, drugs, and our habitats."

A more meaningful definition has been offered by Drs.

Nicholas Ashford and Claudia Miller in *Chemical Exposures; Low Levels and High Stakes:* "[MCS] is an acquired disorder characterized by recurrent symptoms, referable to multiple organ systems, occurring in response to demonstrable exposure to many chemically unrelated compounds at doses far below those established in the general population to cause harmful effects. No single widely accepted test of physiologic function can be shown to correlate with symptoms."

As these twin definitions suggest, MCS is characterized by an intolerance to small amounts of one or more common chemicals in the environment. Said more plainly, patients suffer reactions that are similar to, but distinctly different from, allergic responses when in the presence of even minute quantities of chemical compounds for reasons that remain unclear. Perfumes, synthetic fabrics, plastics, solvents, and other pollution sources barely noticeable to the average person are often unbearable to people with MCS, much the way pollen is intolerable to hay fever sufferers.

Most MCS patients fall into one of two categories: those who have become ill as a result of chronic exposure to low levels of chemicals and those who have been exposed to high levels of toxicants over a short period of time. Some research indicates a handful of common substances may have the potential to trigger MCS by making exposed individuals sensitive to extremely low levels of chemicals or to other toxic substances. Chief suspects in this class include formaldehyde, pesticides, chemical solvents, mercury compounds, acrylic resins, and isocyanates.

Links to Indoor Air Pollution

No one really knows how many people worldwide suffer from MCS. The most widely accepted estimate comes from the National Academy of Sciences, which in 1987 suggested 15 percent of the U.S. population—some 37.5 million people—may experience varying degrees of environmentally related illnesses and sensitivity to chemicals found in common household products, including pesticides, rubber, solvents, detergents, metals, and other substances.

But environmental medicine specialists, who are sometimes called clinical ecologists, believe the NAS estimate understates the prevalence of MCS because physicians often fail to recognize the symptoms of MCS, which can masquerade as other illnesses.

Dr. Ross, a forty-three-year-old conventionally trained physician who specializes in family medicine and allergic disease, says experts have been observing a sharp rise in the number of people claiming to be chemically sensitive in the past decade. Ross says that rise has been reflected in the number of patients seeking care at the Environmental Health Center, which has seen more than 20,000 patients since its founding in the late 1970s by Dr. William Rea.

Interestingly, Ross notes that the increase in MCS cases has mirrored a rise in traditional allergic disease and asthma in the past decade. In the United States, for instance, the number of people afflicted with allergic rhinitis—hay fever—has risen to twenty-two million, while asthma rates have doubled over the past ten years.

Echoing other MCS specialists, Ross credits two factors with the emergence of MCS and the rise in allergic disease: increasing indoor air pollution problems in homes, schools, offices, and other buildings as a result of energy efficiency measures begun in the 1970s, and the nation's expanding reliance on chemicals and chemically treated products since World War II.

"One of the reasons more and more people are reporting chemical sensitivities and allergies is we are constantly polluting ourselves," he says. "You don't have to be a rocket scientist to understand we are increasingly exposed to a polluted environment. I think MCS is emerging now [as] a reflection of increasing pollution in the air, food, water, and our own personal environment."

Ross believes indoor air pollution is playing "a very big role" in the emergence of MCS. "It's far more important, in my opinion, than outdoor air pollution," he says. "We're constantly exposed to chemicals in the indoor environment. And sometimes the levels of indoor pollution are ten to one hundred times the levels of outdoor air pollution. When you look at the studies EPA and others have done on volatile organic compounds (VOCs), the indoor levels are quite alarming. The frightening thing is you're breathing it in all the time."

"I don't mean to sound like I'm saying the sky is falling, but I think we have to recognize the ramifications of indoor air pollution. We've traditionally spent a lot more time and effort on outdoor pollutants, but we're exposed to indoor pollution almost all the time."

Signs and Symptoms of MCS

Tracking MCS has proved a difficult affair for doctors and researchers alike, largely because the signs and symptoms of the disease can be numerous and complex. MCS sufferers have been known to experience an astonishing array of physical symptoms, running the gamut from mild irritations to life-threatening conditions.

According to the scientific literature, manifestations of MCS can include such relatively simple things as skin rashes, watery eyes, sore throats, ear infections, feelings of weakness, and hoarseness. But it can also have serious impacts on the body's immune defenses and other organ systems. Frequently, the respiratory system is involved—evidenced by chronic colds, coughing, wheezing, postnasal drip, asthma, sinusitis, laryngitis, and bronchitis. Gastrointestinal and urinary problems have been observed—abdominal pains, nausea, bloating, diarrhea, constipation, colitis, cramps, vaginal discharge, and back pain. The vascular system can be affected—tension headaches, nosebleeds, hemoptysis, spontaneous bruising, and acne. And the brain and nervous system are often targets—headaches, anxiety, fatigue, concentration difficulties, irritability, disorientation, memory loss, and depression.

Many MCS patients also display other unique characteristics often missed by conventional physicians, according to Ross and Rea. Typically, they cannot tolerate some drugs and medications. Many are allergic to certain foods. They can usually detect extraordinarily subtle odors—natural gas or fabric fragrances, for example—that most people miss. Perfumes, air fresheners, and aerosol sprays may produce nausea or vomiting in MCS sufferers. Many are intolerant of cigarette smoke at any level. And even changes in the weather can make some feel compromised, particularly during smoggy summer months and overcast days when air pollution is held close to the ground.

Because most MCS patients present a mixture of symptoms and odd characteristics, they are often diagnosed by conventional physicians as hypochondriacs and directed to psychiatrists or psychologists. Such diagnoses, which do nothing to help MCS patients, have been legitimized by a handful of widely reported studies that have concluded most people who claim to have environmental illness are actually suffering from commonly recognized psychiatric problems.

In 1985, for instance, a California Medical Association task force study contended: "No convincing evidence was found that patients treated by clinical ecologists have unique recognizable syndromes, that the diagnostic tests employed are efficacious and reliable, or that the treatments used are effective." Four years later, Dr. Ephraim Kahn, a toxic-hazard consultant for the California Department of Health Services, echoed that view in a 1989 article in *The Amicus Journal*: "Most of the symptoms clinical ecologist refer to and treat are not due to any chemical exposure. Rather, they are psychological in nature." More recently, in a 1990 study published in *The Journal of the American Medical Association* (*JAMA*), Dr. Donald Black, a psychiatrist at the University of Iowa Medical College, concluded most MCS patients suffer from commonly recognized psychiatric disorders, such as classic depression and anxiety, that may explain part or all of their symptoms. And as recently as December 1992, the American Medial Association's Council on Scientific Affairs reiterated that point of view in a *JAMA* article that concluded, "No scientific evidence supports the contention that [MCS] is a significant cause of disease or that the diagnostic tests and the treatments used have any therapeutic value."

In addition, MCS has been viewed skeptically by the American College of Physicians and the American Academy of Allergy and Immunology, whose position paper on the topic argues, "The existence of an environmental illness as presented in clinical ecology theory must be questioned."

In the face of such skepticism, Ross and his colleagues remain stoic and philosophical.

"I think it's reasonable to have a healthy degree of skepticism," Ross says. "I myself was conventionally trained as a physician and was skeptical at first. But after I developed chemical sensitivity, it changed my perspective a lot. When someone is exposed to diesel exhaust or eats the wrong food and finds his muscles ache, or if he gets befuddled or has cloudiness of thought on exposure to chemicals, and that's reproducible, then it reinforces the reality of the problem."

"I understand where some physicians are coming from who have not been trained in this at all. Physicians are high on the scale in society, and they come to think they know everything—perhaps more than they do. And when a patient doesn't fit a recognizable diagnosis, the physician begins to say there must be a psycho-social problem here. All too often I think that because we don't understand something we relegate it to a psychological problem."

"That's precisely what happened with sick building syndrome, isn't it? Because it wasn't a phenomenon that was understood at the beginning, it was said to be mass hysteria and psychological in nature. Of course, we now know that's [false]."

Ross acknowledges some of the patients who come to the Environmental Health Center for treatment *are* actually suffering from psychological or psychosomatic ailments. And, where appropriate, they are referred for psychiatric care. But Ross says those patients comprise only 5 percent of the total number of patients who visit the Dallas center. "Having come from the conventional side of things," he says, "I don't think my ability to recognize psychiatric problems is lacking, and we will refer patients to psychiatrists where it's appropriate."

One of the reasons Ross and other MCS specialists are so confident of their diagnoses is increasing evidence of distinct differences in the body functions and organ systems of MCS patients that may be signatures of the illness. For instance, MCS sufferers often have depressed white blood cell counts. Their immune systems are often compromised, with T-lymphocytes, commonly called T-cells, measuring below 1,000—well below the normal 1,260 to 2,650 range. And there is increasing evidence that the brain function of MCS patients is markedly different from that of other individuals.

According to a forthcoming study jointly conducted by the Environmental Health Center and a Dallas aerobic medicine clinic, MCS patients subjected to so-called SPECT examinations, which provide a snapshot of the brain's metabolic function, display abnormal cerebral defects similar to those commonly seen in habitual drug abusers.

"What we're finding," Ross says, "is that the MCS patients with neuro-cognitive kinds of symptoms have patches of irregular brain activity. It suggests there is a dysfunction in the metabolic activity of the brain that is different from psychiatric disease and dementia. It's a neurotoxic, brain injury type of pattern like we see in a drug abuser."

One of the more intriguing findings of this still preliminary research, Ross notes, is the SPECT scans often show improvements in the metabolic activity of the brains of MCS patients after they go through a "detox" program by eliminating chemicals from their diet and reducing contaminants in their home and work environments.

More Support

In spite of the strides made in recent years in understanding the many facets of MCS, environmental medicine specialists continue to face scorn and criticism from many in the mainstream medical community. Still, there is evidence that a growing number of public health and environmental specialists are falling in line behind MCS experts. Research conducted recently at Harvard, MIT, Yale, Johns Hopkins, and other universities has provided increasing support for a physiologic basis for MCS. In addition, a number of scientific and government organizations are adopting supportive views. Among them:

The NAS. In estimating that 15 percent of the U.S. population suffers from MCS, the NAS in 1987 recommended an eighteen-month study of the problem, convened a panel in 1989 to examine the relationships of toxic exposures and immune system response, and conducted an MCS workshop in 1991, at EPA's request, to develop research protocols for the syndrome. In addition, the private nonprofit society of scientists and researchers released a book on MCS in 1992 whose introduction by Jonathan Samet and Devra Lee Davis noted: "The emergence of [MCS] as a phenomenon that needs investigation coincides with recognition that myriad exposures to environmental agents are sustained in indoor and outdoor environments. Regulatory activity has focused on outdoor air pollution, but children and adults . . . spend most of their time indoors and personal exposures to many air pollutants are thus determined largely by indoor pollutant concentrations. Monitoring studies conducted during the 1970s and 1980s showed that homes, public buildings and industrial workplaces can be contaminated by diverse gaseous and particulate pollutants that originate in unvented combustion, evaporation of volatile agents from various materials and solutions, grinding and abrasion, soil gas and biologic sources. Absorption of the pollutants can occur through the lungs, gastrointestinal tract and skin."

EPA. Echoing the NAS, EPA officials have produced several documents on indoor air pollution that explicitly recognize MCS as a health problem meriting investigation. In its Total Exposure Assessment Methodology (TEAM) study of volatile organic compounds in the 1980s, the agency called MCS a major public health risk among individuals chronically exposed to low levels of chemicals indoors. In a 1988 EPA indoor air guidebook, the agency advised homeowners and building managers to reduce occupant exposures to formaldehyde and other chemical substances indoors, noting "there is some evidence that some people can develop chemical sensitivity after exposure." And in its 1989 "Report to Congress on Indoor Air Quality," the agency again singled out MCS as a serious health problem, noting illness can result from levels of chemical exposure well below federal regulations and industry guidelines.

Congress. Legislation pending before both branches of Congress cites MCS as a health problem associated with indoor air pollution and calls for increased federal efforts to research and address the risks. The Indoor Air Pollution Act sponsored by Rep. Joseph P. Kennedy II (D-Mass.) asserts "a portion of the population of the United States may have heightened sensitivity to chemicals and related substances found in the air indoors" and directs EPA to investigate risks to sensitive subpopulations. A less sweeping Senate bill sponsored by Sen. George Mitchell (D-Maine) also acknowledges "indoor air pollutants pose serious threats to public health (including cancer, respiratory illness, multiple chemical sensitivities, skin and eye irritation, and related effects)" and provides funding for research.

In addition, Ashford and Miller, in their exhaustive review of the subject, *Chemical Exposures: Low Levels and High Stakes*, offered the following assessment in 1991:

"The acceptance of chemical sensitivity as a bona fide physical disease may also be facilitated by the recognition that it is widespread in nature and not limited to what some observers would describe as malingering workers, hysterical housewives, and workers experiencing mass psychogenic illness. We are struck by the fact that individuals in such demographically divergent areas as industrial workers, housewives, and children report similar polysymptomatic complaints triggered by chemical exposures."

What's Going On?

The average person is bombarded with hundreds, if not thousands, of toxic chemicals and hazardous substances every day in virtually every facet of modern life. Research by the Consumer Product Safety Commission (CPSC) shows the air inside the average American home can contain up to 150 different chemicals from household products and other sources. EPA studies of American homes indicate chemical levels tend to be two to five times higher indoors than outdoors. And research by the National Institute for Occupational Safety and Health (NIOSH) and the EPA has found that new energy-efficient office buildings can trap hundreds of hazardous substances, including dozens of carcinogens.

How is it, then, that the average person can metabolize these substances without any apparent difficulty, while the chemically sensitive patient cannot?

That question is at the center of the controversy and debate over MCS. While theories abound about the origins of the illness, no clear explanation for MCS has emerged. "We don't yet know the mechanism of MCS," acknowledges Randolph. "But empirically and pragmatically, there is no question about the cause and effect demonstrable in these afflicted persons."

In the absence of scientific certainty, MCS specialists have advanced several theories that might explain what's happening in the development of MCS. One holds that the human body can only handle so much exposure to toxic chemicals before becoming overloaded. Using a simple analogy to describe this complex physiological process, Ross explains: "Some people can metabolize chemicals reasonably well; some can't. It's almost as if we are like sponges—we can only soak up so much chemical exposure before something has to give."

The concept of a "total body load" is important in understanding MCS, he says. The combined effect of all chemicals and pollutants taken into the body—components of contaminated air and water, pesticides in foods, and viruses, molds, bacteria, and pollens—can distort body functions. MCS patients have been found to have significant body loads of such chemicals as organochlorine pesticides, volatile organics, solvents, polychlorinated biphenyls (PCBs), and herbicides.

In addition to the combined effects of chemicals in food and work and home environments, specialists believe genetic factors may also play a part in individual susceptibility to MCS. Recently, the Environmental Health Center conducted a case-control study of twenty-five chemically sensitive patients undergoing "detoxification" programs—avoiding chemicals and contaminants in their diets and environments. The group was compared to a control group of twenty-five Mormons, whose chemical exposures are generally lower than those among the general population because of religious restrictions on drugs, alcohol, tea, coffee, and other foods. Biological tests on the Mormons turned up traces of relatively few common contaminants, including industrial pollutants and car exhaust, but similar tests performed on the MCS patients turned up a much wider variety of chemical contaminants, despite the fact that they had for years been limiting their pollutant exposures to levels far below those experienced by the Mormons.

This study suggested that MCS patients may have a latent inability to purge contaminants from the body, making them susceptible.

"What I think explains the findings is that they have a genetically predetermined detoxification ability," Ross says. "That is, they have a tendency to collect contaminants and be more susceptible to chemical exposure."

In addition, Ross notes, MCS patients are frequently deficient in vitamins and minerals involved in the body's natural processes of eliminating toxins, including Vitamins E and C, beta carotene, and selenium. That deficiency may be partly genetic and partly due to nutritional factors.

Adaptation

A second fundamental MCS concept is a principle known as "adaptation" or "masking." In the first stages of MCS, patients may actually build up a tolerance to a particular substance by "adapting" to it, thus noticing no ill effects. To diagnose MCS, specialists must then attempt to "unmask" or "de-adapt" an afflicted individual by having the person avoid the suspected incitant for three to five days. Finally, they "rechallenge" the patient by reintroducing him to the offending substance in a carefully controlled environment. In a person who is truly chemically sensitive, re-exposure to such a substance will often produce an obvious, stronger physical reaction after a period of nonexposure.

Experts believe this interplay between adaptation and unmasking is a continual, unrecognized process that many industrial and non-industrial workers unwittingly experience routinely. Painters and battery workers, for instance, often notice the substances they handle on the job may bother them after returning from a long vacation, until they get used to them again. The phenomenon may also explain why reformed smokers—who "adapted" to tobacco while smoking then "de-adapted" after they quit—are often most offended by secondhand smoke.

The mechanisms that cause some people to develop MCS in this way are still not entirely understood. Some researchers have theorized that environmental pollutants may damage the body's natural immune systems and enzyme detoxification systems. Ashford and Miller have proposed an additional hypothesis: chemicals may harm the brain and nervous system in such a way that the body no longer distinguishes between low and high levels of chemical exposures.

What most experts appear to agree on, however, is that MCS is quite different from traditional allergic reactions, where the body dispatches antibodies to attack specific foreign substances, such as ragweed, pollen, and animal dander.

"What we're talking about here is a two-step process," explains Ashford, an MIT professor who has extensively studied MCS. "People seem to complain about an event—a chlorine spill or a pesticide exposure or moving into a new energy-tight building—after which they become exquisitely sensitive to a number of substances of similar or dissimilar chemical class.

"What happens then, we believe, is the brain misreads subsequent [low] exposures to chemicals as large exposures and sends the body in all kinds of flight-or-fight responses, including inappropriate signals to the immune and/or endocrine systems."

Interestingly, Ashford and others have also identified several sensitive subpopulations that may be at greater risk of MCS, based on reviews of the literature on exposure. Among them:

• Industrial workers, who experience chronic and acute exposure to industrial chemicals (primarily blue collar men, twenty to sixty-five years old)

• Occupants of "tight buildings," including office workers and schoolchildren who are exposed to off-gassing chemicals from construction materials, office equipment and supplies, tobacco smoke, and inadequate ventilation

• Residents of communities whose air or water is contaminated from toxic waste sites, pesticide spraying, or local industry (all ages, male and female, middle to lower class)

• Individuals who have had personal and unique exposures to various chemicals, pesticides, drugs, and consumer products (white middle- to upper-middle-class professionals)

Ashford regards the emergence of these sensitive groups as perhaps the most important link to unraveling the mysteries of MCS. "It's the fact that these groups are so different and so well defined that, in our view, in the future will give the clue to what this problem is and how it unveils."

Treatment

As scientific researchers continue to study the possible mechanisms at work in MCS, the Environmental Health Center has carved out a unique niche in the practical treatment of chemically sensitive patients since 1974. The center's staff includes a full complement of physicians, medical assistants, nutritionists, technicians, counselors, nurses, and support staff totaling sixty in all. The center includes a specially designed "environmental unit" for outpatient care and utilizes a nearby research laboratory for blood testing and other work. It also has its own non-toxic condominium complex—which fetches $40 to $50 per day, $280 to $350 per week—for patients visiting the center from out of town.

One of the center's distinguishing features is its unconventional design. Everything inside the complex has been carefully fashioned to offer MCS patients a veritable refuge, or "oasis," from a polluted world. The walls and ceilings are made of smooth porcelain on steel, giving the place the appearance of a huge bathtub. The floors are terrazzo tiles with non-toxic grout. Sophisticated air-purification systems, using HEPA (high-efficiency particle air) and charcoal filters, keep the air free of contaminants, chemicals, pollens, molds, and dust throughout the facility. Most furniture is made of wood or steel and all fabrics are untreated and made from natural fibers. Bottled water is available, and tap water is specially filtered. Employees refrain from using scented perfumes, soaps, and other personal care items. Even toxic compounds from the computer system are controlled with a specially made stainless-steel housing that encloses all wires and electronic components, allowing the air-filtration system to scavenge contaminants from off-gassing plastics and equipment and vent them outdoors.

The chief difference between diagnoses at the Environmental Health Center and those at conventional medical clinics is that the center's staff includes an assessment of environmental factors in its analyses. Diagnosis at the center begins, for instance, with a complete medical profile that includes not only a patient's personal and familial history of illness, but also an accounting of chemicals that may be present in one's diet, home, workplace, and consumer products. Physical examinations and laboratory tests typically follow—including blood tests to check for body-toxin levels, vitamin and mineral deficiencies, T- and B-lymphocytes, IgE (immunoglobulin E) antibodies, organ function, and immune system problems. Diagnostic procedures and X-rays are often employed. Skin testing is performed to determine sensitivities. Finally, with difficult patients, the center's staff conducts "challenge" tests in the environmental unit to unmask potential chemical sensitivities. The center, whose services are honored by many insurance carriers, charges $125 for an initial office visit. Follow-up visits are $25 to $50. Patients are also billed for diagnostic tests and consultations.

To describe treatment at the center as "holistic" is to understate the point. In short, it encompasses all factors that influence a patient's health, from environmental controls to nutritional management and supplementation, immunotherapy, stress management, and biofeedback. Boiled down to its most fundamental level, however, the cornerstone of therapy at the center is based on avoidance of chemicals. Patients are taught how to lead less toxic lifestyles and limit their exposures to chemicals and contaminants—at home and work, through diet and medications—that can trigger MCS.

Center staff maintain that once a patient reduces his "total body load" of toxic agents, he can often tolerate small exposures to chemicals and contaminants that would otherwise cause illness or reaction.

Predictably, some quarters of the medical establishment frown on such practices. In its position paper on the topic, the American Academy of Allergy and Immunology sums up the position of the mainstream medical community in this way: "Review of the clinical ecology literature provides inadequate support for the beliefs and practices of clinical ecology. . . . Diagnoses and treatments involve procedures of no proven efficacy."

Ross and his colleagues disagree but accept the slings and arrows of the medical establishment with resignation. Unfortunately, MCS patients are often less able to handle the pressures of being in the middle of one of the most contentious medical debates of the times.

Ross and other specialists also believe increased attention to environmental issues generally, and indoor air pollution risks specifically, can only improve the visibility of MCS and environmental medicine techniques. Already there are signs that scientific researchers and the media are paying more attention to the issues. In 1992, for instance, many of the nation's major-league newspapers devoted extensive coverage to two important developments on the MCS front: reports that the illness may be striking dozens of workers who used chemical cleaning agents to help clean up oil from the 1989 *Exxon Valdez* spill and thousands of Operation Desert Storm veterans who were exposed to oil well fires, pesticides, petrochemicals, spilled crude, and other environmental contaminants during the Persian Gulf War.

"This is really the first pertinent publicity this has had in thirty years, and it's the major cause of the veterans' problems, I have no doubt," says Randolph, who has himself examined at least four Persian Gulf War veterans. "I think this may finally blow it open, so [MCS] will have greater acceptance."

To Ross, these recent developments are both good news and bad. He explains: "I think it is a portent of things to come. We've perhaps worn out the analogy of canaries in the coal mines, but I don't think it's off the mark at all."

"In any situation some people will react more easily than others, but the fact that we're seeing increasing numbers of people affected tells us there is increasing pollution in the world. Unfortunately, that means it's going to get much worse before it gets better."

Echoing the others, Ashford suggests the public health debate on MCS actually encompasses larger environmental health issues, including the roiling policy squabbles over indoor air pollution policy.

"Notice the tremendous regulatory challenge that this presents for the agencies," he observes. "We're talking about people who, once sensitized, are sensitive in the parts-per-billion range. So are we talking about a no-risk, not chemically based economy? Well, that just isn't sensible."

"On the other hand, it's obvious that we haven't controlled pesticides, drugs, and other sensitizing chemicals sufficiently. Perhaps if we begin to control pesticides and other chemicals, then we won't create a new generation of highly sensitive people."

Wimberley

While the debate continues to rage, a small peace is being made in the United States in Wimberley, Texas. It's not uncommon to see people whip out gas masks at the slightest provocation in supermarkets or department stores or along the highways of the 10,000-resident town. The mail is hand-delivered to the community of MCS folks to limit diesel and car fumes in the rural region where they live. Most Wimberley-area residents seem to understand, or at least accept, that Sue Pitman has good reason to sometimes sleep on her porch. Or that her neighbor, Dot Dimitry, "detoxifies" her mail by hanging it on a clothesline in the open air before reading it. Or that Larry Martin washes his clothes—all white cotton—in hydrogen peroxide. Or that Janet Bennett spends most of her days locked up in her porcelain-walled trailer.

"Wimberley is a very special place," says Sue Pitman. "Lots of people have moved here to get away from the chemicals, so the idea that people can react to chemicals is not such an unusual idea for a lot of the people who live around here.

"Most people are supportive, some aren't. It's a small town, and our situation challenges the status quo. The people that have an investment in the status quo have a problem with us and they snub us. On the other hand, there are other people here who are unbelievably supportive. And the merchants are generally very friendly. But the good old boys constantly make jokes."

Sue says she is troubled less and less by the fact that the MCS patients are often viewed skeptically by the public at large and by many practitioners of conventional medicine.

"You just have to accept that certain people can't understand," she says. "But it's not all in our heads. When you see it happen to your children, when you can see the cause and effect time after time after time, you just can't deny it.

"I've been working for a long time to just open people's eyes to the possibility. I understand the skepticism and I could talk forever about why some doctors feel that it's not real. I didn't even believe Dr. Randolph when he first suggested this to me. I had to go through and make the lifestyle changes we made to see that it really made a difference in our lives."

Sue Pitman also believes that what is happening to the people in Wimberley has implications that go far beyond the prairie town's limits.

"I think that we're the canaries, and people better pay attention because we see it happening all around us, and it seems to be happening more frequently," she maintains. "As the world becomes more and more permeated with synthetic chemicals that our bodies are not equipped to deal with, we're going to keep poisoning ourselves and our children.

"The thing is, we have the power to change our lives, and it's so easy to live a toxic-free lifestyle."

CHAPTER 2

A Guide to Common Indoor Air Pollutants

In its promotional literature, the Bionaire Corporation maintains its air filters are among the most effective available—removing up to 99 percent of potentially harmful particles, fibers, dusts, and bacteria from the air. But for sheer air-purification capacity, the advanced air-cleaning technologies developed by Bionaire and other air-filter manufacturers can't possibly hold a candle to a single pair of human lungs.

You don't need a degree in respiratory biology to recognize that the lung is a truly remarkable piece of equipment. Consider:

• Each pair of lungs contains 300 to 400 million alveolar sacs, which absorb oxygen and other gases into the bloodstream to be distributed to vital organs. If the alveoli from an average set of human lungs could be unbundled from the chest and spread out over a flat surface, they would cover an area roughly the size of a regulation tennis court.

• The average person inhales about 20,000 liters of air each day, or about 10 each minute, 600 each hour. That's sixty pounds of air flowing through the average person's lungs each day—a volume far greater than the two liters of water and food that enter our bodies daily.

• The lungs have remarkably sensitive and sophisticated mechanisms for filtering and responding to the complex mixture of gases, particles, and foreign substances contained in a single breath of air. Even the cleanest air contains not only life-sustaining oxygen, along with nitrogen, carbon dioxide, and water vapor, but also hundreds of potentially harmful particles and trace elements—dust, pollen, microorganisms, industrial pollutants, auto exhaust components, smoke, and chemicals (both natural and manmade). It's been estimated that the average person inhales up to two heaping tablespoons of assorted particles and contaminants each day. Most are filtered out by nasal hairs and the cilia of the lungs and trachea, which work together with a mucous "escalatory" system to allow impurities to be excreted through sneezing, coughing, and blowing the nose. But in individuals exposed to high pollution levels or particularly noxious agents, these respiratory systems can be overwhelmed. And just as oxygen is absorbed through the alveolar walls of the lungs—one-millionth of an inch thick, in places—so can harmful agents penetrate sensitive lung tissues. It takes just one microscopic grain of pollen to trigger a reaction in allergic individuals, just one microscopic tuberculosis bacillus in fifty cubic meters of air to cause pulmonary tuberculosis.

Taken together, these factors demonstrate why the lungs are the body's front-line defense against many forms of pollution. They also indicate why the lungs are so vulnerable to toxic agents and why the connection between health and the environment has historically centered on air quality and respiratory disease.

"Breathing is our biggest interaction with the environment and the primary way that toxins enter the body," observes Dr. Joseph Brain, chairman of the Harvard School of Public Health's Department of Environmental Health. "It's also something we're going to have to do wherever we are. And because the air in a variety of indoor, outdoor, and work environments is frequently—sometimes intrinsically—contaminated, the structure of the lung makes it vulnerable."

Historical Links Between Health and Air Pollution

Some of the earliest accounts linking air pollution and health problems emerged during the Roman Empire, which some historians believe may have fallen because lead in the water pipes and culinary utensils destroyed the brains of its leaders.

In the United States, the first comprehensive municipal air pollution laws were passed in 1881 in Chicago and Cincinnati, where furnace smoke was declared a public nuisance. Thirty years later, state legislation passed in Boston was the first to acknowledge air pollution as a regional and national phenomenon.

But it took several serious, well-documented air pollution episodes to convince the U.S. Congress that airborne contaminants posed grave enough risks to require uniform federal standards, first enacted in 1955 with the Air Pollution Control Act. Historians largely credit two incidents with forcing Congress's hand in 1955. The first was a 1948 episode in Donora, Pennsylvania, where twenty people died and hundreds fell ill after high concentrations of pollutants from local iron and steel plants were held close to the ground by a thermal atmospheric inversion. The second, in 1952, was a far more serious episode that left some four thousand people dead from domestic coal-burning facilities during a similar inversion in London.

Today, of course, the link between air pollution and human health is no longer in doubt. And increasingly stringent health-based federal environmental laws have been generally credited with driving small to moderate reductions in the six most serious outdoor pollutants—lead, carbon monoxide, nitrogen oxides, sulfur dioxides, and particulate matter and ozone smog—for which the EPA has set ambient air quality standards.

Still, there is a preponderance of evidence that suggests that air pollution–related health risks have never been so great.

For all of the progress on outdoor air pollution, for instance, 164 million Americans lived in cities and counties where outdoor air pollution levels exceeded at least one federal ambient air quality standard in 1991, according to a recent study by the Centers for Disease Control and Prevention (CDC). What's more, the federal government has barely scratched the surface on air pollution problems indoors, which may pose even greater public health risks that remain largely unrecognized by the public. "While improvements have been made in our outdoor air quality in recent years, more studies are finding that Americans could be at substantial risk from exposure to indoor air pollutants," noted American Lung Association President Dr. Lee B. Reichman, commenting on a 1993 Gallup poll showing only 24 percent of Americans believe indoor air pollution is a serious problem in the home. "We feel many Americans may fail to recognize the serious lung health effects of indoor air pollutants such as radon, environmental tobacco smoke, mold, and bacteria."

In addition to these troubling findings on air quality, U.S. health statistics on a handful of diseases traditionally associated with air pollution and environmental factors have not been encouraging. According to federal health experts, the United States is experiencing a rise in both the rate and severity of asthma, allergies, and certain cancers.

Fighting the War on Cancer

Since 1971, when the U.S. Congress passed the National

Cancer Act to combat cancer, the nation has spent more than $22 billion to improve patient treatment, enhance early detection methods, and expand research. But while some progress has been made in specific areas—improving survival rates for some childhood cancers and for patients under sixty-five—overall death rates from cancer are actually on the rise. In 1991, the General Accounting Office of Congress soberly noted that while some gains in treatment have been made, "We must conclude that there has been no progress in preventing the disease."

It might seem paradoxical that as the average life expectancy in the United States has risen, more Americans are dying today of cancer. Yet, according to the American Cancer Society, there has been a steady increase in the cancer-mortality rate over the last half-century. In 1930, the age-adjusted rate was 143 people per 100,000 population; today it is about 170 per 100,000. In addition, the ACS predicts eighty-three million Americans now living will eventually have cancer. That's one in three. Nearly two million will be informed by a physician this year alone that they have some form of cancer. And at least fourteen hundred Americans die of the disease each day.

In addition, cancer is now the leading cause of death for American women and second, behind heart disease, among men. This is largely due to the incidence of lung cancer, which increased among women by 2 percent between 1973 and 1987 and is the leading cause of cancer death for both sexes in the United States. If current trends continue, the society projects cancer will overtake heart disease to become the leading killer of men and women by the year 2000, with breast, prostate, and kidney cancer and melanoma skin cancer all rising.

In addition to the staggering body count suggested by these harsh statistics, cancer has been estimated by the National Institutes of Health's National Cancer Institute to cost the nation $104 billion a year—a third of it in direct medical costs.

While genetic factors clearly play some role in the onset and development of cancer, the ACS noted in its 1992 "Cancer Facts

& Figures Report" that the primary factors driving these sobering cancer statistics are environmental in nature.

"Most cancer cases in the United States are believed to be environmentally related, that is, associated in some way with our physical surroundings, personal habits, or lifestyles," according to the report's authors.

"Some environmental causes are well known. About 30 percent of all cancer deaths are directly related to the use of tobacco. Most skin cancers result from ultraviolet radiation."

"Other causes are harder to assess. Diet is suspected as an important element in cancer risk, some say causing perhaps 35 percent of all cancers. . . . Various occupational hazards, especially ionizing radiation and chemicals like asbestos, benzene, and vinyl chloride are known to cause cancer when exposure levels are high. Overall, however, workplace exposures account for only a small percentage of all cancers."

Asthma

In 1980, fewer than four of every 300,000 Americans died from asthma, a respiratory disease that causes abnormal constriction of the branching tubes that carry air into the lungs. By 1990, however, the figure had leapt 46 percent to nearly 6 in 300,000, with some four thousand asthma deaths a year, according to the CDC. The problem, federal officials say, appears to be particularly acute for blacks, whose death rates increased by 52 percent—from 2.5 deaths per 100,000 to 3.8 per 100,000.

In addition, children seem to be afflicted at much higher rates of the ailment, with federal statistics showing 4.3 percent (or 2.7 million) of children under eighteen years old suffered from asthma in 1988, compared to 3.2 percent in 1981. Those figures translate to ten million missed days of school, 200,000 hospitalizations and nearly thirteen million doctors visits a year, according to the CDC. The disease is also more common in women than in men, with federal statistics showing twice as many women than men aged forty-five to sixty-four suffer from asthma, which now strikes about one in twenty people.

No one is certain just why the rate has skyrocketed. Certainly some of the growing caseload is due to better diagnosis of asthma. Some researchers suspect outdoor and indoor air pollution, which have been shown to trigger asthma attacks, may be key factors. This argument is bolstered by statistics showing higher rates among minorities and urban dwellers, who may live in apartment buildings where insects, mold spores, dust mites, and other triggering indoor allergens proliferate.

"The major categories of triggers for asthma are in the air we breathe: air pollution and inhaled allergens," notes Dr. Stephanie Shore, a lung biologist at the Harvard School of Public Health, explaining that cold air, stress, and exercise can also bring on attacks. "In terms of the kinds of pollutants that can trigger asthma, we can see they are not just in the outdoor but also in the indoor air, where we spend most of our time."

"Cigarette smoke, smoke from woodburning stoves, fumes from cleaning products, sprays, and other products in the home have also been shown to trigger asthma. Allergens are found both in the outdoor air—things like ragweed pollen—and also indoors."

Some of the most unassuming things can also be the worst causes of asthma. Mold in poorly ventilated kitchens and bathrooms can trigger asthma, the family pet, dust mites in bedding, furniture, and carpeting. What's more, research has found many of these asthma triggers can also induce hyper-responsive reactions even in healthy lungs if levels are high enough or present for long periods of time. Many air pollutants and allergens can result in an influx of white blood cells into the airways, causing them to become inflamed. In addition, children in homes with smokers are also much more likely to develop asthma, suggesting that constant exposure to smoke-borne irritants can initiate the disease. Viruses, particularly in small children or infants, have also been shown to cause asthma by invading the cells of lung airways and altering the genetic material that determines their function. It has also been established that occupational exposures to such things as cotton dust, grain dust, and chemicals can cause asthma.

Asthma is best managed by routine monitoring and/or drugs, which can cost $50 to $100 a month. But the nation's ten million asthma sufferers must also usually control their indoor environments by reducing exposure to allergens and pollutants.

Allergies

Similar trends have been observed in traditional allergic disease since the 1970s, which experts say has experienced an unprecedented rise that is fueling new concerns about indoor and outdoor air pollution problems. Federal statistics indicate about forty million Americans suffer from some form of allergy, which results from aberrant functioning of the human immune system that causes it to "attack" harmless foreign substances mistakenly perceived as a threat to the body.

Historically, allergy research and outreach has centered on the immune system's response to outdoor air pollutants and allergens, such as pollens from pollinating trees, grasses, ragweed, and other plants that trigger reactions in the nation's twenty-two million hay fever sufferers. But allergists and researchers have increasingly turned their attention to indoor air pollutants as the primary triggers in many cases of allergic reactions as modern construction practices have reduced ventilation in homes, schools, offices, and buildings.

For instance, droppings from the lowly dust mite—a microscopic insect that thrives in house dust, bedding, carpets, and furniture—have been found to trigger reactions in as many as 50 percent of people with allergies. Molds, mildew, and fungi, present in virtually every home, are also believed to be a major cause of allergic reactions. And as many as two million Americans are believed to suffer from allergies to cat dander.

Public Health Implications

In light of these and other health trends, many in the medical community are turning ever more attention to the role of the environment, and particularly indoor air pollution, in the nation's public health picture. At the same time, policy specialists are placing an increasing emphasis on disease *prevention* over *treatment* as the most cost-effective and reliable means of addressing pressing public health concerns.

This trend is almost certain to continue with the graying of the baby boom generation and increasing advances in the ability of environmental engineers to detect contaminants in the air and environment. Thirty years ago, for instance, chemical concentrations could barely be detected at parts-per-thousand levels; today sophisticated gas chromatography and mass spectrometry offer the ability to detect the presence of some substances in the parts-per-trillion range.

"The 1990s, I think, are becoming the decade of the environment," observes Dr. Brain of Harvard. "It's absolutely clear that the environment does affect health and that there is disease and death being caused by the environment."

Taking Action on Indoor Air Pollution

Amid all the discouraging statistics on environmental pollution and disease, there is some cause for hope. We know, for instance, that while many of the diseases associated with pollution exposures cannot be cured by medicine—cancer, asthma, and allergies among them—they can often be prevented through prudently controlling environmental exposures. Thanks to the efforts of leading public and private environmental health researchers, there is a significant body of information about how to reduce human exposures to toxic contaminants and risks of disease, often through simple, inexpensive means.

This is especially true of indoor air pollution exposures,

which can almost always be more readily controlled than those outdoors through individual actions taken in one's own home, office, and workplace. "We know of many hazardous situations we can control," Dr. Brain adds. "We're concerned about indoor radon, occupational exposures, indoor cooking, cleaning, allergens. And we know that some of the things we're doing, like insulating our homes better, are leading to fewer air exchanges and more indoor air problems. But the good news is we're making a lot of progress and we know how to address many of these problems."

With these facts in mind, the passages that follow attempt to summarize and detail the risks and sources of the most significant indoor air pollutants. They also present risk-reduction strategies advised by public health officials and EPA experts for combating indoor air pollution in homes, offices, schools, workplaces, and other buildings. The intent is not so much to detail technical solutions for indoor air quality problems, which often require professional assistance, but to provide information about the prevalence and risks of some common contaminants.

Environmental Tobacco Smoke

In his last major news conference as EPA administrator, William K. Reilly ended his four-year stint with the Bush Administration with an astonishing admission. In one of his final official acts as the EPA's top dog, Reilly said he thought it "odd" that the federal agency had spent most of its energy and federal funding regulating environmental problems that pose small public health risks and far too little time targeting bigger threats.

The occasion, appropriately enough, was a press conference called in January 1993 to release the findings of EPA's most comprehensive analysis of the risks of environmental tobacco smoke (ETS)—a report Reilly himself deemed "as important and influential" as any in the EPA's twenty-three-year history.

The peer-reviewed study, four years in the making, was the first to declare ETS a "Group A" human carcinogen—a designation given to only ten other potent hazardous compounds, including arsenic, benzene, asbestos, and radon. It also estimated that ETS is responsible for some 3,000 lung cancer cases and another 300,000 lower respiratory-tract infections among children each year.

For most of the one hundred-plus reporters present, the news conference was an anti-climactic conclusion to two years of rancorous debate over the EPA's findings on ETS, first aired in draft copies of the 530-page report leaked to the press in 1990. The only thing that had held up its release was pressure from the tobacco industry, which had contended the EPA's methodology was faulty and its risk-assessment projections flawed.

Yet Reilly, clearly hoping to end his service under Bush with a bang, managed to turn the press conference into something more than just a rehash of old news on the hazards of ETS. Less than a minute into his prepared remarks, Reilly made it clear that he hoped the EPA report would become the foundation for increased limits on workplace smoking and more aggressive efforts on indoor air pollution. In an unusual digression, he even took a few moments to urge the environmental health reporters gathered in the EPA Education Center to spend more time analyzing indoor air problems, contending that radon and ETS pose risks of greater magnitude than virtually any other risk, chemical or otherwise, regulated by the federal government.

"In my opening statement before the Senate at my confirmation hearing four years ago, I said I thought it odd that EPA and the nation had invested so much priority and money in cleaning up the air outdoors, and had paid hardly any attention to the problem of air pollution indoors, where people spend 90 percent of their time," Reilly said. "I pledged to increase the priority for indoor air quality. . . .

"The report I am releasing today I expect to be as important and influential, both in the United States and throughout the

37

world, as any EPA has ever done. Because of the large percentage of time we all spend indoors, ETS and radon may present the most important environmental health risks we face today. And ETS in particular is something we personally can take steps to prevent, especially in the home, where we spend so much time, and in the confined spaces of automobiles, which dramatically increase concentrations."

Robert Axelrad, who heads the EPA's Indoor Air Division, underscored Reilly's points after the press conference. "I think the tobacco issue is one of the most important, if not the most important, indoor air pollution issues," he said. "But we don't want people to believe addressing ETS is addressing indoor air pollution." Eric Butthauer, the EPA's director of research under Reilly, also said the ETS report would provide increased emphasis on indoor air quality (IAQ) issues at the EPA. "It's become very clear that indoor air is a major concern for the health of Americans," he said, noting EPA is developing strategies for ranking and managing IAQ risks. "But what we don't have is a database that is really convincing. When we reach that point over the next several years, I think you'll see that concern grow, just as it has with ETS."

Where There's Smoke . . .

Despite the two-year delay in its release, the EPA's 1993 report on ETS has come to be regarded as the most comprehensive analysis of the risks of passive smoking ever conducted. History will almost certainly place the report on par with the 1964 Surgeon General's report declaring cigarette smoking a health hazard and the 1986 Surgeon General's report implicating ETS in lung cancer.

Specifically, the 1993 EPA report examined more than thirty major studies of the links between ETS and cancer, and more than one hundred studies of ETS and childhood respiratory health. Its conclusions were scrutinized and endorsed by the agency's Science Advisory Board, an independent panel of scientific experts and ETS specialists. That panel, which included two researchers whose works have been funded by tobacco industry interests, conducted public hearings on the report and examined more than one hundred studies cited by critics of the EPA's conclusions.

Among the report's findings:

• Smoking is the leading cause of preventable death in the United States, with smokers accounting for 25 percent of the population. Every year, 425,000 Americans die as a result of diseases related to smoking—140,000 from lung cancer alone.

• ETS is responsible for approximately 3,000 lung cancer deaths among adult non-smoking Americans each year.

• ETS causes serious respiratory problems for children and is responsible for between 150,000 and 300,000 cases of lower respiratory tract infections, such as bronchitis and pneumonia, among children up to eighteen months of age. More than 7,500 of these ailments result in hospitalization.

• ETS increases the frequency of episodes and severity of symptoms in asthmatic children, worsening the conditions of between 200,000 and one million asthmatics every year. It is also a potential cause of new asthma cases in healthy, asymptomatic children—perhaps causing as many as 26,000 new cases a year.

• ETS increases the prevalence of fluid in the middle ear, a sign of chronic middle ear disease, the most frequent cause of hospitalization in young children for an operation. Consequently, this condition imposes a heavy financial burden on the health care system.

• ETS irritates the upper respiratory tract and is linked to a small but significant reduction in lung function.

• ETS contains more than 4,000 substances—43 of which are known or believed to cause cancer. What's more, many are present at higher levels in sidestream smoke curling up from the ends of lit cigarettes than in the mainstream smoke actually inhaled by smokers. The implication is that sidestream smoke may account for the biggest tobacco risks to smokers and non-smokers alike.

These conclusions echoed the findings of the 1986 Surgeon General's report, which in analyzing eighteen studies found ETS causes lung cancer in non-smokers, raises the number of respiratory infections in children, and slows lung growth in youngsters.

As with the Surgeon General's findings, the EPA's 1993 conclusions were immediately challenged by the Tobacco Institute, an industry trade group that argues that the cancer risks associated with smoking have been largely exaggerated. But the EPA conclusions were based on several primary analytic findings the SAB deemed appropriate and difficult to challenge scientifically.

Among those findings was that ETS and mainstream smoke inhaled by smokers are similar in chemical and physical composition; that mainstream smoke is widely accepted to be a potent human carcinogen; and that ETS physically permeates the bodies of non-smokers exposed to tobacco smoke, with carbon monoxide, nicotine, and other "biomarkers" detected in their bodily fluids. In addition, the thirty-plus studies on which EPA based its conclusions—focused on people who never smoked, usually spouses of smokers—all found strong associations between ETS and lung cancer.

A Dose of Perspective

One of the most provocative conclusions of the ETS report was a section comparing the health risks posed by ETS to other environmental contaminants. Based on the study's conclusions, the EPA projected a smoker's risk of developing lung cancer at about 1 in 10 to 1 in 20, while people who have never smoked face a lung cancer risk that is many times lower—about 1 in 200. But according to the EPA's projections, 20 percent of all lung cancers in non-smokers are due to ETS. That means a non-smoker exposed to low-grade ETS faces at least a 1-in-1,000 risk of developing lung cancer, while individuals chronically exposed to high levels of ETS (spouses of smokers, for instance) face a 2-in-1,000 risk.

Risks in this range, while seemingly small, are considered quite high by health experts. By comparison, the EPA generally sets its standards and regulations such that risks are below 1 in 100,000 to 1 in 1,000,000. In other words, the risks associated with ETS are at least an order of magnitude greater than those posed by any other hazard the EPA currently regulates.

In announcing the ETS findings, Reilly noted they would have no immediate practical impact on the EPA, which has no authority to regulate indoor air pollutants. But Reilly said he expected the EPA document would drive the Occupational Safety and Health Administration (OSHA), which is charged with regulating workplace hazards of all kinds, to act. He also predicted an avalanche of new state and local laws to restrict smoking, as well as increased bans on smoking in the private workplace—predictions which have largely proved true.

"With this report we have laid the firm scientific foundation upon which policy can now be built," he added. "While EPA has no regulatory authority over ETS, we will be working with [OSHA], which can regulate smoking in the workplace. I would hope that this report will form the scientific basis for a policy on their part."

Where the Rubber Hits the Road

As it turns out, the EPA report on ETS has actually had a greater impact on OSHA than on the EPA itself. Under federal law, OSHA is required to compel private employers to provide workplaces that limit "recognized hazards" to the lowest "feasible levels." Consequently, within days of the EPA report's release, former Labor Secretary Lynn Martin asked OSHA officials to provide a detailed list of options for dealing with ETS, which, as of this writing, remains at the top of the list of

OSHA's workplace safety priorities.

For OSHA, the EPA report came at a critical juncture in the agency's own deliberations on indoor air quality policy. Since 1991, the agency has been assessing the need for indoor air regulations and soliciting comment from all corners in the public policy debate. That process, which had not progressed beyond the "request for information" phase as of 1993, was altered somewhat in the wake of the EPA report on secondhand smoke to allow OSHA to examine proposals to expand ETS restrictions in the workplace independently of its comprehensive IAQ analyses.

But even as OSHA deliberations continue, health advocates predict the federal government will have no choice but to limit smoking in the workplace now that the EPA has recognized ETS as a significant hazard. The only question is whether such limits will be issued by the Clinton Administration or mandated by the federal courts, where anti-smoking lobbyists have filed suit to force OSHA's hand in the matter.

"In light of federal laws requiring OSHA to protect workers from recognized hazards, they will have to ban it," contends Kathleen Scheg, legal counsel for the anti-smoking group Action on Smoking and Health, which is seeking a court order for OSHA to ban workplace smoking. "To do anything else would be a neglect of responsibility."

Food for Thought

Even before the ink was dry on the EPA press releases, the Department of Health and Human Services, through the CDC, prepared an extensive public education program—including targeted print, television, and radio ads—to bring the EPA report findings to a wider audience. Three of the campaign's television public service announcements noted that workers are 34 percent more likely to develop lung cancer if exposed to ETS on the job, eleven thousand children are hospitalized each year for exposure to ETS, and the air inside a smoke-filled bar can be six times more polluted than the air near a busy highway. Another ad admonished: "Tobacco is the only consumer product that, when used as directed, causes death."

Like Reilly, former Health and Human Services Secretary Dr. Louis W. Sullivan had been an outspoken advocate of workplace smoking limits throughout his tenure under President Bush. The EPA and HHS were, in fact, the only two federal agencies that had banned smoking in their buildings. Sullivan used the announcement of the EPA report to unveil findings of a sweeping study by HHS's National Center for Health Statistics, completed too late to be included in the EPA report, that found that infants are three times more likely to die of sudden infant death syndrome (SIDS) if their mothers smoke during and after pregnancy. Sounding an alarm echoed by advocates of stronger indoor air regulations, Sullivan observed, "Our children have no control over the conditions under which they live, attend school, or frequent public places. Therefore, parents, school officials, business owners, public health policy makers—indeed, all of us—bear a special responsibility to protect our children from this menace to their health and well-being."

Citing health studies showing some nine million American children under the age of five live in homes with at least one smoker, Sullivan committed his agency to working for bans on smoking at all schools, the passage of clean indoor air laws, and expanded workplace smoking limits by the year 2000. He also took a few shots at the tobacco industry, noting it spends $4 billion a year—$500,000 an hour—on advertising to increase cigarette sales, which in 1989 netted the industry $7.2 billion in after-tax profits.

While the public-policy impacts of the EPA report and the HHS informational campaign are still being felt, there were some early and definitive responses in a handful of states and industries. Beginning in places like Florida, Massachusetts, and California, languishing workplace smoking efforts were invigorated almost immediately. The EPA also quickly prepared a workplace smoking guide for private employers, calling for expanded restrictions. And even the International Board of Governors of the Building Owners and Managers Association (BOMA), representing real estate industry interests, voted unanimously to endorse a federal workplace smoking ban within weeks of the EPA report. Today, forty-six states and the District of Columbia in the United States plus many places in other countries now have regulations on the books that in some way ban or restrict smoking in public places, according to the HHS's Public Health Service.

Still, progress has remained slow on several fronts. As of this writing, no comprehensive federal policy has emerged on smoking in public places or at the workplace. And the laws adopted by many states and communities vary widely in the absence of federal standards. Some are tough, some are not. Much, therefore, remains left to individual homeowners, workers, employers, and building managers.

Fortunately, the HHS and the EPA have increased the flow of public information on the risks of ETS, including a series of recommendations for reducing exposure to secondhand smoke in homes, workplaces, restaurants, and other public places. What follows is a brief summary of those guidelines.

Risk Reduction Strategies

At Home

Controlling ETS in the home is the single most important step one can take to limit exposure to indoor pollution. Limiting smoking indoors also saves on time and money spent on cleaning curtains, walls, and windows and may bring lower insurance rates in some regions. To keep a smoke-free home:

• Ask visitors not to light up, and have gum, mints, and foods available as alternatives. Those who cannot abstain can be shown outdoors or to a porch.

• Encourage any smokers living in the home to quit, smoke outdoors, or at least confine the habit to a single well-ventilated room closed off from the rest of the house.

• Observe these same practices while driving or riding in a car, and encourage schools, daycare centers, and other facilities frequented by children to go smoke-free.

In Restaurants

Even restaurants with smoke-free sections can't eliminate ETS risks, because ventilation systems tend to recirculate the indoor air and don't filter out tobacco smoke. Consequently, as the HHS notes, trying to have a smoke-free section of a restaurant is like trying to have a chlorine-free section of a swimming pool. To limit exposure to ETS in restaurants:

• Encourage a favorite restaurant to become smoke-free by appealing to the owner or manager, noting that cleaner air makes food taste and smell better.

• Point out that three-quarters of the American restaurant-going public do not smoke, most non-smokers find ETS annoying, and 90 percent of non-smokers ask to be seated in non-smoking sections.

• Explain that tobacco interests often operate behind restaurant association front groups that argue smoking bans will hurt business. There is no evidence from any city that has passed a 100-percent smoke-free restaurant ordinance that such measures are detrimental to business. Some restaurants even draw in new customers eager to escape smoky environs elsewhere.

• Finally, use the power of the pocketbook: Patronize smoke-free establishments and encourage friends to follow suit.

In the Workplace

In the wake of the EPA report, many employers are taking steps to limit ETS exposures and their own liability as federal, state, and local laws are being developed on workplace smoking. In general, the only way to reduce ETS is to make buildings smoke-free or limit smoking to separately ventilated areas. Advice for workers seeking ETS limits:

• Express employee concerns to the boss or manager. Stress the fact that a smoking policy would solve an existing problem, not create a new one, and would send a message that the company is concerned about the health and well-being of its employees.

• Remind managers and employers that removal of a known health hazard from the workplace protects companies from possible lawsuits from employees affected by ETS. Point out that ETS is responsible for thirty times as many lung cancer deaths among non-smokers as all regulated air pollutants combined and that there is no known safe level of exposure to a carcinogen.

• Note CDC studies that have found a smoker typically costs his or her employer at least $1,000 more per year in increased health care costs and decreased productivity than a non-smoker. Private studies have suggested total costs may reach as high as $5,000 per year per smoker. Company life, health, and fire insurance premiums are often lower for employers with smoke-free workplaces; encourage benefits managers to investigate competitive insurance plans that offer such breaks. Note that computer equipment, furniture, carpets, and other furnishings last longer in non-smoking environments.

Carbon Monoxide and Other Combustion Gases

On a bitterly cold evening three weeks before Christmas 1988, John and Linda Cifarelli tucked their two daughters into bed, fired up the gas boiler in their split-level home, and turned in for the night. It was the first time they were to use the boiler, which fired a driveway de-icing system, since the family had moved into their home the previous September.

Tragically, it would also be the last.

While the family slept, carbon monoxide from the boiler crept into every room in the house, due to a faulty exhaust system. By morning, John and Linda were dead, along with their twenty-three-month-old daughter, Nina. The couple's youngest child, five-month-old Annabelle, survived, as did a house guest who opened his bedroom window after nausea and a headache woke him. Medical tests later showed that John, who was thirty-four years old, and his twenty-six-year-old wife, several months pregnant at the time, had high levels of carbon monoxide concentrations—between 60 and 75 percent—in their blood at the time of death. Young Nina also had a carbon monoxide concentration well above 50 percent. Annabelle's was far lower—6 percent—possibly due to her lower body metabolism or because a blanket may have covered her face while she slept. The house guest, Andrew Csermak, had a carbon monoxide level in the 20-percent range.

The Cifarelli case, while tragic, is not unique. According to the CDC, carbon monoxide accounts for about one-half of all accidental poisoning deaths in the United States, killing at least 1,800 Americans each year. In addition, carbon monoxide disables another 10,000 and is the deadly culprit in more than 2,300 American suicides annually. In most cases, carbon monoxide poisonings are the result of improperly vented combustion gases from indoor gas furnaces, space heaters, grills, stoves, and cars in attached garages.

Until recently, victims of carbon monoxide poisoning were thought primarily to be among the poor, who sometimes use faulty space heaters or gas stoves to heat their homes. But the trend toward tighter, super-insulated homes and a rise in the use of gas-fired heating systems—now in half of the nation's 110 million homes, according to the Department of Energy—have increased the number of carbon monoxide poisonings among more well-off homeowners in recent years.

As high as the CDC figures are, the statistics reflect only

those cases where causes of death have been definitively established by measurements of carboxyhemoglobin—a "biomarker" found in the blood of carbon monoxide–poisoned victims. And because carbon monoxide is virtually undetectable in the air—it is odorless, colorless, and tasteless—some experts believe the actual number of fatal carbon monoxide poisonings is far greater than currently estimated.

One of the reasons so many carbon monoxide poisonings occur each year is because the invisible gas is produced by a wide variety of common indoor sources. Another is that the sources and hazards of carbon monoxide are not generally known as well as might be expected. In January 1990, for instance, the British journal *Lancet* reported the bizarre case of a family whose members complained to hospital officials about nausea, abdominal pain, and headaches that appeared to have no apparent cause. Upon learning family members had been drinking unrefrigerated milk, doctors diagnosed family members with infective gastroenteritis and sent them home. But several hours later, four members—including one who had steered clear of the milk—returned with similar symptoms. A subsequent visit to the home revealed the family was using an unvented gas grill to cook meals in the living room.

The case was certainly unusual. But it spotlights the fact that most victims of carbon monoxide poisoning have little or no knowledge of the dangers of carbon monoxide—often until it is too late.

Sources of Carbon Monoxide

In many ways, carbon monoxide is the indoor pollutant of greatest concern among health experts and indoor air quality specialists because it is so pervasive in indoor environments, and at high concentrations, can kill within minutes. Carbon monoxide is produced by the incomplete combustion of gas, oil, wood, natural gas and other fossil fuels. Consequently, it is emitted by a staggering array of sources.

Gas ranges are a chief source of indoor carbon monoxide—particularly older models with gas pilot lights that burn continuously. Kerosene or gas space heaters—used by an estimated eight million people, according to the CPSC—are also common emitters of carbon monoxide. In addition, furnaces, fireplaces, wood stoves, gas dryers, water heaters, and garaged cars can all emit high levels of carbon monoxide indoors if improperly vented, operated, or maintained.

EPA assessment studies have shown households are the primary contributors to personal carbon monoxide exposures. Cars and workplaces, especially commercial and public buildings with attached or underground garages, are also chief areas of concern.

Health Effects of Carbon Monoxide

Inhaled and absorbed through the lungs, carbon monoxide works its deadly business by binding with hemoglobin in the bloodstream and displacing oxygen. In effect, carbon monoxide squeezes out oxygen in the blood and reduces its ability to deliver oxygen to vital tissues and organs such as the brain, heart, and nervous system.

Although there are no uniform health-based standards for indoor levels of carbon monoxide, the federal ambient outdoor limit set by the Clean Air Act is nine parts per million over an eight-hour period—a standard still not being met in forty one major metropolitan areas in the United States, according to the EPA. In setting the carbon monoxide standard, the EPA aimed to keep the average person's blood–carbon monoxide level under 2 percent, the level considered safe for most people.

Health effects from carbon monoxide poisoning often masquerade as cold and flu symptoms—headache, dizziness, drowsiness, disorientation, nausea, abdominal pains, vomiting, eye irritation, and loss of muscle control. As a result, physicians often misdiagnose carbon monoxide poisoning. At high doses, carbon monoxide can lead to brain damage, coma, and death. Although most studies have proven inconclusive about carbon monoxide at low levels, some have suggested chronic carbon monoxide exposure may also accelerate atherosclerotic symptoms and increase risks for thromboembolism in the heart or brain.

People with pre-existing medical conditions, including cardiovascular disease and blood disorders, are especially vulnerable to carbon monoxide exposures, as are developing fetuses, young children, and seniors. And because carbon monoxide blocks the blood's oxygen-carrying capacity, health officials generally suggest persons exercising near carbon monoxide sources may also be at risk. In other words, joggers who run near highways during rush hour may actually be doing themselves more harm than good.

Although health effects vary, a person's condition generally worsens with continued exposure to carbon monoxide. In healthy individuals, concentrations of 3 to 5 percent (a level common among cigarette smokers) can cause minor effects and drowsiness. At concentrations of 10 percent (the level experienced by a firefighter near a blaze scene), nausea and headaches become more pronounced. At 15 percent (a traffic cop at a busy intersection), blurred vision, severe nausea, and pounding headaches are common. At 25 to 30 percent, the average person will lose consciousness. The fatal range is generally considered anything above 40 percent. In unhealthy individuals, seniors, or young children, however, even a 15 percent concentration can bring on coma, and death can result from readings as low as 35 percent.

Other Combustion Pollutants

In addition to carbon monoxide, two other indoor combustion gases—sulfur dioxide, to which asthmatics are extremely sensitive, and nitrogen dioxide—have also been shown to damage the human respiratory system and lead to impaired breathing and chronic bronchitis. In addition, respirable particulates from wood stoves, fireplaces, heaters, cigarette smoke, and other combustion sources can pose significant risks in indoor environments.

While these combustion gases and particles are not generally believed to be highly toxic at very low levels, they are pervasive in homes and buildings and can lead to eye, nose, and throat irritation, chronic lung disease, heart trouble, and other ailments if left unchecked.

Sulfur Dioxide

EPA studies have found that as many as ninety-six million people in America alone may be exposed to sulfur dioxide emissions from gas stoves for an average of four hours per day. "Only a subset of these individuals are expected to be at risk," the EPA concluded but added, "The extent of their risk is highly uncertain."[1]

Nitrogen Dioxide

EPA and World Health Organization (WHO) studies in the 1980s found that children chronically exposed to even low levels of nitrogen dioxide and other combustion gases tend to have higher rates of respiratory symptoms and illness than others.[2] Other clinical studies have determined nitrogen dioxide can cause substantial changes in pulmonary function in normal, healthy adults at concentrations as low as 2 parts per million, and trigger reactions in asthmatics at levels of 0.5 ppm.

Particles

Respirable particles are dangerous because they can include such things as toxic trace metals and carcinogenic polyaromatic hydrocarbons (PAHs). But they can also pose secondary risks by allowing other chemical contaminants, which attach to them, to be inhaled deep into the lungs.

According to the American Lung Association, the incidence of lung disease due to combustion gases and particles is on the rise because people are closing up their homes with insulation and other means to save energy and cut down on heating bills.

Risk-Reduction Strategies

As with other indoor contaminants, the keys to reducing exposures to carbon monoxide and other combustion products involve checking sources for emissions and ensuring proper ventilation. Among the specific strategies experts advise are the following.

• Periodically have professionals check heating equipment, furnaces, water heaters, appliances, and other devices that produce combustion gases. EPA studies indicate that carbon monoxide concentrations due to faulty appliances can easily rise to between 100 and 200 parts per million, well above the agency's 35 ppm one-hour outdoor carbon monoxide standard. Trouble spots include venting systems that can "backdraft"—venting combustion gases indoors—as well as furnace heat exchangers, fan motors, belts, air filters, and fuel-shutoff switches. Consider replacing a very old furnace with a safer, more energy-efficient model.

• Fireplace chimneys, furnace flues, and wood stove exhausts can be blocked by sooty buildup, fallen leaves, and bird nests and should be checked annually and cleaned regularly.

• Never use a gas stove for home heating purposes. When buying a new one, be sure it is equipped with an electric-spark ignition system and not a continuously burning gas pilot light.

• Install an exhaust hood over the kitchen stove that is vented outdoors, and use it whenever the stove is on. Also check to be sure burner flames are blue and not yellow-orange, as this indicates the stove is not burning properly.

• Don't use kerosene heaters in enclosed spaces at home or

while camping, and never use gas or charcoal grills indoors.

• Consider installing a carbon monoxide detector (cost: $25 to $250), which functions like a smoke detector.

• Avoid sitting in a parked car with the engine running for long periods of time. Never allow a vehicle to idle while in an attached garage. Don't tailgate in stop-and-go traffic. EPA studies have found that carbon monoxide levels can quickly rise to 60 parts per million—nearly twice the agency's federal outdoor one-hour standard—inside cars stopped in traffic jams.

• Inspect car exhaust systems, including mufflers and tailpipes, for carbon monoxide leaks.

• When renovating property, making additions, or insulating living space, make sure there is adequate ventilation.

Radon: The Nation's Number One Environmental Health Risk

In 1984, Stanley Watras became a troubling historical footnote of the Atomic Age when he inexplicably set off the radiation-checkpoint alarms at the Limerick Nuclear Power Plant in Pennsylvania—while on his way *into* the facility. Initially, reactor technicians couldn't make sense of it because Watras's job as a plant engineer required no contact with radioactive materials. But when officials checked Watras's home, they found the culprit was not the Limerick reactor at all, but radiation from radon gas seeping into his home and being carried into the plant on the worker's clothes.

Subsequent tests determined Watras's home was built on top of a uranium-rich geologic formation known as the Reading Prong that cuts through the Northeastern United States. Consequently, the Watras family was being bathed in radon levels nearly one thousand times higher than those deemed safe under federal guidelines, posing risks roughly equivalent to smoking hundreds of cigarettes a day and receiving more than half a million chest X-rays a year.

Stanley Watras's experience was among the first widely reported public health alarms on radon, an odorless, colorless gas present at some level in virtually every home. Although the risks of radon had been well documented by the scientific community in the 1970s, it wasn't until the Watras case drew national headlines that federal regulators began assessing the wider public health risks of the naturally occurring gas.

Since 1987, the EPA has consistently ranked indoor radon the nation's number-one environmental health problem, estimating it causes 14,000 lung cancer deaths a year, with a possible range of 7,000 to 30,000. That makes radon, classed as a Group A human carcinogen, the second-leading cause of lung cancer behind tobacco smoking. It also makes it the most dangerous of all indoor air carcinogens, with federal health statistics showing that radon-induced lung cancer kills more people each year than fires, drownings, and airline crashes combined.

Radon is a decay product of radium and uranium, present in nearly all rocks and soils on Earth, including granite, shale, and phosphate. As a result, it is by far the biggest source of human exposure to radiation, accounting for 55 percent of our total exposure, according to the National Council on Radiation Protection and Measurements. (By contrast, nuclear power accounts for less than 1 percent; medical X-rays, 11 percent.)

Outdoor exposure to radon is not believed to pose significant public health risks. But in the indoor environment, where Americans spend most of their time, radon gas can be trapped in well-insulated homes and buildings and rise to dangerously high levels. Radon itself is actually an inert gas, but with a half-life of about four days, it decays into several harmful radon progeny, sometimes called "radon daughters." Two of those radon decay products, polonium-238 and polonium-214, emit radioactive alpha particles that, when inhaled, can damage lung tissue and lead to cancer.

EPA risk projections of the public health effects of radon stem from studies of underground miners. As early as the sixteenth century, high rates of respiratory disease were observed in underground miners, according to the National Research Council of the National Academy of Sciences.[3] Although some scientists have challenged the EPA's radon risk projections, the agency's radon assessments are stronger than those concerning most other regulated environmental contaminants because they are based on studies of human miner populations.

The EPA has no statutory authority to regulate radon, but the agency has jointly published a "Citizen's Guide to Radon" with the U.S. Department of Health and Human Services. First issued in 1986 and updated in 1992, the guide outlines radon risks and provides guidance for action when concentrations exceed the agency's designated "safe" level. To date, the EPA has urged all American homes and schools to be tested for radon. In addition, the agency issued a "Home Buyer's and Seller's Guide to Radon" in 1993, urging radon tests be performed at the time of any real-estate transaction.

Unfortunately, the message about radon doesn't appear to be getting out. The EPA estimates only 6 percent of the nation's 110 million homes have been tested for radon, while a recent national poll by the Roper organization found Americans rated radon the second-lowest of twenty-nine health risks.

Sources of Radon

Concentrations of radon in indoor air are expressed as picocuries per liter (pCi/L) of air, a measure of the rate of decay. (A piocurie is one-trillionth of a curie, a standard measure of radiation that takes its name from Marie Curie.) The average indoor radon level has been estimated to be 1.3 pCi/L, with outdoor levels found in the 0.4 pCi/L range. Although the EPA believes no level of radon is entirely safe, agency guidelines recommend homes with levels of 4.0 pCi/L or higher to be modified to reduce radon concentrations.

According to EPA projections, nearly one out of every

fifteen homes exceed the 4.0 pCi/L cutoff. But in some regions of the country—the Northeast, for instance—up to one in five homes may have elevated levels. All told, the EPA estimates between eight million and ten million homes have unsafe levels of radon.

Although radon can be found in private well water and some building materials, the principal source of radon is soil gases seeping into a home from cracks, holes in foundation walls and floors, drains, sumps, pipes, and other openings. Consequently, radon tends to pose less of a threat in upper floors of apartment buildings and commercial structures than in free-standing single-family homes. At greatest risk are tightly-insulated and poorly ventilated homes, where "negative" air pressure can act as a vacuum and draw radon in through foundation gaps.

Reducing Risks from Radon

The good news about radon is that it is easy to detect and not too expensive to address. The EPA and other health officials offer the following advice on radon testing and abatement:

• To determine the level of radon indoors, first conduct a "short-term test," using a unit that typically remains in the home for two to ninety days. Such devices are inexpensive (less than $25 for most charcoal-canister detectors) and available at hardware stores.

• When buying a radon-testing device, make sure it is state certified or has passed the EPA's testing program.

• When using the test, place it in the lowest level of a home used as living space (i.e. the basement if it is frequently used, otherwise the first floor), but not a bathroom or kitchen. Windows and doors to the area should be closed at least 12 hours before starting the test and throughout the test period. Devices should be placed at least 20 inches from the floor and exterior walls, and in an area free from drafts, high heat, and humidity.

• If a short-term test detects a radon level of 4.0 pCi/L or higher, follow it up with a second short-term device or a long-term unit such as an alpha-track detector to confirm the results. If that second test also turns up a high reading, steps should be taken to reduce the radon levels.

• If conducting a radon test while buying a home, be wary of radon tampering, which occurs in as many as 40 percent of real estate transactions, according to the American Association of Radon Scientists and Technologists. Ask whomever is conducting the test what tampering safeguards will be used and consider a follow-up test to be sure readings are consistent.

• High indoor radon levels can be reduced by sealing cracks in foundation walls and floors, but this method alone does not significantly lower concentrations. The EPA's method of choice for abating radon is "sub-slab depressurization"—a system of pipes and fans that can be installed without major changes to most homes. These systems, which typically cost $500 to $2,500, remove radon gas from below the foundation and safely vent it to the outside. Many new homes include such systems as a feature.

• Other common abatement methods involve blockwall suction systems that remove radon from homes with hollow spaces in concrete foundations; heat-recovery units that replace indoor air with fresh outdoor air; and draintile-suction techniques that link up water drainage systems with basement air exhaust units.

• When hiring a radon abatement contractor, be sure to use one certified under the EPA's Radon Contractor Proficiency Program, which trains and tests licensed professionals. Lists of qualified professionals can be obtained from the EPA or state officials.

Lead: Children at Risk

Lisa and Michael Griffin thought they knew all there was to know about lead poisoning. Soon after their fourteen-month-old daughter, Ashley, was diagnosed with high blood-lead levels in

1989, they moved from their lead paint-ridden Boston apartment into a nearby two-family home owned by Lisa's mother, who had deleaded the property shortly after her granddaughter's diagnosis. But in 1992, after Ashley had undergone chelation to reduce the lead in her bloodstream, the Griffins' family doctor once again informed Lisa and Michael that high levels of lead had been found in Ashley's blood. This time around, Ashley's baby brother, seventeen-month-old Sean, also was diagnosed with lead poisoning. The culprit: peeling lead-based exterior paint from a neighbor's home that was carried into the backyard and sandbox used by the children, according to a lawsuit filed by the Griffins in 1993.

"We thought we did everything we could to protect them," says Lisa Griffin, who is twenty-five years old, noting that both her children now have learning disabilities caused by their lead exposures. "It's tough, it really is. My daughter [now four] is in special needs classes, and Sean still isn't speaking [at thirty months]. And the house next door still isn't fixed because they have no children. . . ."

Horror stories like this one are far more common than one might think. According to the CDC, between three million and four million children under age six have lead levels in their bloodstream high enough to cause learning disabilities, lowered intelligence and behavioral problems. That makes lead poisoning the biggest environmental health problem facing children.

Until recently, lead poisoning was thought to be primarily a problem for the nation's urban poor, with unsupervised children at greatest risk of poisoning from ingesting flaking chips of lead-based paint. But federal studies have found lead poisoning can result from ingesting minute particles of lead present in house dust and soil, not just from eating sweet-tasting lead paint chips. And while Department of Health and Human Services statistics indicate poorer, urban minorities are at greatest risk (55 percent of all lead-poisoned kids come from impoverished black families), lead poisoning is becoming increasingly common among middle- and upper-income folks living in suburban areas.

As the Griffins' case demonstrates, the health consequences of lead poisoning are taking a tremendous human toll. But the financial costs of the problem also loom large on the horizon. The CDC estimates that the United States spends $30,000 for every case of lead poisoning for medical, educational, and other societal costs. In addition, some experts have estimated it may cost $200 billion to remove lead from all American homes, schools, and other buildings where it is present. That's thirty times the EPA's annual budget and twice what American industries spend each year to comply with *all* federal environmental laws. Currently, however, the federal government spends less than $250 million a year on the problem.[4]

In 1992, growing concerns about the risks of lead drove Congress to pass two new laws that will increase the federal government's involvement in targeting lead problems. The first, part of the Preventive Health Amendments of 1992, directs the CDC to increase grant moneys to state and local agencies to screen children for elevated blood-lead levels. The second, the Residential Lead-Based Paint Hazard Reduction Act, requires home sellers by 1995 to disclose to prospective buyers the presence of lead in homes built before 1978 and directs the EPA to set health-based standards for lead in paint, soils, and dust.

While these initiatives are expected to increase the federal government's role on such issues in the long run, comprehensive lead policies and programs will still be left to individual cities and states to write and implement in the short term. For most states, this means that homeowners, tenants, landlords, and real estate professionals will continue to be on their own in dealing with the hazards of lead.

Screening for Lead

In the absence of comprehensive federal lead regulations, at least thirty-five states in the United States and places in other countries have passed some type of lead-related statute aimed at reducing the risks from household lead paint, toys, lead batteries, and occupational settings, according to the ad hoc Alliance to End Childhood Lead Poisoning. Of those, sixteen have mandatory blood-lead screening laws requiring preschoolers to be checked for high lead levels by pediatricians and clinics.

In Massachusetts, which passed the nation's first lead law in 1971 and currently boasts the most stringent lead paint regulations, children under six years old are systematically screened for the presence of lead in blood. Elevated lead levels trigger a series of follow-up medical actions, source identification programs, and mandatory remediation efforts designed to reduce risks to family, neighbors, and friends. Massachusetts and several other states have also established programs to license lead abatement contractors and provide income tax credits for low- and middle-income residents who delead their properties.

Massachusetts law also requires all homes with children under six to be deleaded, but this mandate has inspired heated criticism from real estate interests who say the statute has done more to block real estate transactions and strap landlords than it has to address lead paint problems. The reason, critics note, is that it costs almost $3,500 to delead the average five-room residence, with costs reaching upwards of $10,000 for larger homes, according to Massachusetts records.

The National Association of Home Builders and the National Association of Realtors have also been critical of mandatory lead removal laws and have urged state and local governments to focus on lead education and in-place management of paint risks. Consequently, many states are investigating the use of paint-like "encapsulants" that seal in lead paint and reduce risks of chipping, dusting, and cracking. Such products, which are now manufactured by about two dozen companies, offer the availability of reducing lead paint risks at a significant savings over more conventional abatement methods. To date, however, they have been used in only a few pilot programs in Maryland,

Pennsylvania, and some federal housing projects.

In 1993, the state of Massachusetts took the first tentative steps toward more widely licensing and authorizing the use of such products, some of which are fortified with rubber, plastics, fiberglass, and even cement-like reinforcements that resist impacts and bond with even underlying layers of lead paint. Other states, watching Massachusetts's lead law developments closely, are expected to follow suit in coming years.

"The EPA is supporting our efforts here, and there is the sense that, if our program is approved, this could be the model for new national standards on encapsulants," notes Brad Prenney, head of the Massachusetts Department of Public Health's Childhood Lead Poisoning Prevention Program.

"There are still some difficult scientific and technical issues," he adds. "The last thing we want to do is develop regulations and allow products that are going to lead to more problems down the road. But I think everybody would like to see a cheaper way to deal with this problem."

Old Problem, New Interest

Interest in the use of encapsulants has peaked in recent years as concerns about the risks of lead to children have soared. The hazards of lead paint have been known since at least the turn of the century, with many European countries banning its use in the 1920s. But it wasn't until 1978 that the EPA banned the use of lead in nearly all paints made and used in the United States. By that time, the U.S. Public Health Service estimates, varying levels of lead paint had made its way onto the walls, woodwork, and windows of nearly fifty-seven million American homes and three-quarters of all houses built before 1980.

The primary health risk posed by lead exposure is damage to the brain and nervous system. Even tiny amounts of the toxic heavy metal have been shown to cause learning disabilities, lowered IQ, behavioral disorders, speech impairments, and a variety of other neuropsychological problems. Lead exposure has also been linked to kidney and liver damage, reproductive problems, high blood pressure, and anemia.

While workers in certain occupations—the construction industry, for example—are vulnerable to lead exposure, children are at greatest risk from the toxic effects of lead. This is because their young bodies absorb more lead per pound of body weight than do adults. They are also more likely to have nutritional deficits that increase lead absorption, and their developing nervous systems are exquisitely sensitive to lead.

The U.S. Agency for Toxic Substances and Disease Registry has estimated that as many as one in six of the nation's twenty-two million preschoolers have high blood-lead levels. Lead-based paints are the chief source of lead exposure to children, but lead can also be present in house dust, soil, drinking water, food (from lead solder used in canned products), culinary utensils, and foreign-made toys and household products.

Perhaps the most distressing news about lead is that new research increasingly shows that even minute traces of lead can cause health problems in children. In fact, the long-term effects of lead are so devastating to young children that in 1992 the CDC lowered its threshold definition of dangerous blood-lead levels from 25 micrograms of lead per deciliter of blood to 10. That's about 1 part per million in weight. And some health experts believe even this new level is not stringent enough to protect children from the risks of chronic lead poisoning.

Dr. Herbert L. Needleman, a research scientist at the University of Pittsburgh who advises the U.S. government on lead issues, has repeatedly said there is no level of lead in children that has been shown to be safe.

Symptoms

Once ingested or inhaled as specks of lead-based paint or fine particles of soil or dust, lead enters the bloodstream and blocks the production of hemoglobin, which red cells use to carry oxygen to vital organs, and inactivates enzymes in the brain and nervous system. The silvery metal can also be stored in soft tissues and bones for years, only to spill into the bloodstream later in life. That makes it particularly dangerous in pregnant or nursing mothers. According to federal studies, some 400,000 children are born with toxic levels of lead in their blood.

Acutely lead-poisoned children can exhibit symptoms of depression, hostility, loss of appetite, abdominal pain, and anemia. Unchecked, lead poisoning can lead to seizures, coma, and death. Children who are chronically poisoned, however, may experience only subtle changes in temperament, such as irritability, fatigue, and hyperactivity. Because these symptoms are also common in other illnesses, lead poisoning can often be missed.

Lead poisoning is usually treated through a process known as chelation, whereby drugs that bind to the metal in the bloodstream are used to flush it from the system. Chelation is expensive and painful because it usually requires intravenous injections. As a result, experts say prevention—taking the child out of the lead, before taking the lead out of the child—is the best strategy for addressing lead poisoning.

Where It's Found

According to the U.S. Census Bureau, houses built before 1940 have the highest concentrations of lead paint, making the Northeast and the Midwest the regions with the greatest lead-paint problems. Pre-1980 American housing stock is estimated to contain more than three million tons of lead in paint. A breakdown issued by the Census Bureau in 1990 projected the following percentages of older housing stock in all fifty states:

31 to 40 percent: Iowa, Maine, Massachusetts, Nebraska, New York, Pennsylvania, Rhode Island, and Vermont.

21 to 30 percent: Connecticut, Illinois, Indiana, Kansas,

Michigan, Minnesota, Montana, New Hampshire, New Jersey, North Dakota, Ohio, South Dakota, West Virginia, Wisconsin.

11 to 20 percent: California, Colorado, Delaware, Idaho, Kentucky, Louisiana, Maryland, Missouri, Oklahoma, Oregon, Utah, Washington, Virginia, Wyoming.

0 to 10 percent: Alabama, Alaska, Arizona, Arkansas, Florida, Georgia, Hawaii, Mississippi, Nevada, New Mexico, North Carolina, South Carolina, Tennessee, Texas.

These numbers show that while some regions face more serious lead poisoning risks, no state can dismiss the problem and no one is immune. Even drinking fountains in the U.S. Senate's Capitol Hill office buildings have been found to have lead levels well above federal standards. Ironically, some of the highest readings were found outside the hearing room for the U.S. Senate Committee on Environment and Public Works.

Reducing Lead Poisoning Risks

The good news about lead poisoning is that it is preventable. The EPA, the CDC, and other health agencies have come up with a wide variety of strategies for reducing household exposures to lead—some as simple as good housekeeping.

In the Home

• Test for the presence of lead paint, particularly in homes built before 1978. Lead labs and contractors can be hired to detect its presence through chemical tests—usually sodium with sulfide—or new X-ray fluorescence equipment (costing up to $75 per room). If paint is peeling or flaking, keep children away and cover it temporarily.

• Never attempt a lead-removal or renovation project in a leaded area alone. Do-it-yourselfers can cause far more harm than good, because sanding and chipping paint from walls can pound lead into dust that can be easily ingested and inhaled.

• Contact local, state, or federal officials for information on licensed or otherwise qualified lead abatement professionals, who use respirators and special high-efficiency vacuum equipment to contain lead dust. Typically, professionals remove the paint by scraping it off or chemically removing it from surfaces with movable or vulnerable parts—windows, sashes, sills, stair treads, railings, and door frames. Flat walls where lead paint is intact can be covered safely with paneling, wallboard, or some wallpapers. In some cases, removing lead-painted objects such as windows, doors, or frames is the best way to go.

• Screen children for lead at one year of age and every year or two afterward. Blood-lead tests (costing $25 to $75) should begin at six months for children in high-risk situations.

• Keep in mind that the hand-to-mouth activities of young children make them vulnerable to lead dust that sloughs off as paint on walls, woodwork, windows, and doorframes ages. Wash children's hands frequently and keep mouthable items such as toys, bottles, teddy bears, and pacifiers clean.

• Reduce indoor dust to a minimum by washing window sills, ledges, and floors regularly with a phosphate-based cleaner. Also try to limit lead tracked into the home from soils outdoors.

• Don't underestimate the role of a proper diet; high iron and calcium intake can slow down a child's absorption of lead.

• Test any suspected household sources of lead, including crystal wine glasses, decanters, imported ceramic pottery, wine bottle foils, old toys and furniture, metallic fishing weights, antique pewter, and glass artwork.

• While paint is the primary source of lead poisoning, EPA studies show as many as forty million Americans are also exposed to lead in drinking water. Commercial laboratories typically charge under $50 for tests to determine if lead in tap water exceeds the EPA's action level of 15 parts per billion (0.015 milligrams per liter of water). Risks can be reduced by running the tap water for at least a minute before drinking it and not using hot water (which can dissolve lead in pipes more quickly) for cooking and drinking. Licensed plumbers can determine if a home's plumbing contains lead; municipal records name the plumber who installed water lines in particular homes in building permit files.

Outside the Home

• Test suspect yard soils for lead; levels exceeding 500 to 1,000 parts per million should be removed or covered with six inches of soil or grass turf. Tests typically cost under $30 and may be performed free by town officials.

• Plant gardens away from painted structures and roadways and be sure to check the soil for lead before eating fruits and vegetables, which can sometimes absorb the toxic metal.

• Consider using vinyl siding to cover exterior walls with lead-based paint.

Asbestos: Fibers and Phobias

Of all indoor pollution sources, asbestos has generated the most regulatory attention and controversy. In fact, while the health risks of asbestos exposure have been known for decades, public policy debates continue to rage over the role of the federal government in managing and reducing asbestos risks.

Asbestos is not a single substance but a family of naturally occurring minerals prized for their strength, flexibility, durability, and resistance to combustion and chemicals. The most common variety of asbestos used in the United States is chrysotile, but other forms—crocidolite, tremolite, anthophyllite, and actinolite—have been in use since the turn of the century.

The durability of asbestos, which is Greek for "unquenchable," has been known since at least the Dark Ages. Historical footnotes indicate the Emperor Charlemagne awed guests from rival kingdoms by hurling asbestos tablecloths into the fire, then withdrawing them unsinged to "clean" them.[5] Asbestos was first used commercially in Quebec in the Nineteenth Century, when it was mixed with burlap, pitch, and manila paper to form roofing material. Use of asbestos ballooned in the Twentieth Century, when it was hailed as a miracle building material by the construction industry and used in an array of products, including insulation, fireproofing, paints, plaster, and ceiling and floor tiles.

The first serious questions about the public health risks of asbestos were raised in the United States in the 1960s, although studies conducted as early as the 1920s had linked it to respiratory diseases in workers who were mining, milling, and handling asbestos. Among the well-established health impacts of occupational asbestos exposure are lung cancer, caused by long thin asbestos fibers lodging deep in the lungs; asbestosis, a scarring of the lung tissue that impedes the exchange of oxygen and carbon dioxide and can lead to respiratory or heart failure; mesothelioma, a rare and fatal disease characterized by tumor growth in the linings of the lungs or abdominal cavity; and gastrointestinal cancers.

No one knows how many deaths can be attributed to asbestos exposure, but in 1988, EPA risk analysts said the United States could expect a total of 131,200 asbestos-related deaths between 1985 and 2009, most of them due to lung cancer suffered by asbestos miners, millers, and workers exposed to high levels decades ago, when there were no regulatory controls.

Asbestos Laws

Since the early 1970s, forty-four major federal actions have been taken to regulate asbestos in occupational settings, schools, construction activities, commercial uses, and in air and water pollution emissions. At least forty states in America as well as many other countries have also enacted some form of asbestos law. In addition, a series of highly visible asbestos lawsuits have increased pressure on the asbestos industry. In late 1992, more than 26,000 asbestos cases were pending in the federal courts alone.

Beginning in 1971, the EPA instituted a series of increasingly stringent restrictions on the use of asbestos. In 1973, the agency first banned its use on pipes, boilers, structural beams, and in fireproofing and insulation materials. Since then, the agency has promulgated standards for its abatement in a variety of settings and established a program for accrediting personnel involved in abatement projects in non-residential buildings. In 1986, Congress passed the most wide-ranging federal asbestos law, the Asbestos Hazard Emergency Response Act, which directed the EPA to issue rules requiring all schools to be inspected and freed from asbestos risks through removal or in-place management. Three years later, the EPA announced plans to ban the manufacture, importation, and processing of nearly all asbestos products by 1997. Taking this action under TSCA, EPA officials said, would save up to 202 lives at a cost of some $450 million.

In the wake of these initiatives, federal officials estimate the use of asbestos plummeted from a peak of 560,000 metric tons to under 85,000 metric tons in the late 1980s.[6] Yet, despite this trend, asbestos remains present in many homes, schools, and office buildings and in a staggering array of products. A 1988 EPA study found, for instance, that 733,000 of the 3.6 million public and commercial buildings in the United States—not including schools or small residential buildings—contained asbestos. The survey indicated 501,000 of those buildings contained loose asbestos, which can release fibers into the air, and 317,000 were plagued with "significantly damaged" asbestos posing immediate risks. EPA studies also indicate up to 35,000 American schools—a third of the total—may be exposing some 15 million children and 1.4 million employees to asbestos.

The National School Board Association has estimated that the total cost of removing asbestos from schools where it is present will top $6 billion. To seal off or remove it from all commercial and government buildings could reach $51 billion, the EPA estimates. What is less certain is how much it would cost to abate the risks in the tens of millions of homes believed to contain it.

Debate in Asbestos Handling

In the early days of asbestos regulation, the EPA generally urged the removal of asbestos from buildings as the best way to reduce risks. Since then, however, there has been an evolution in the thinking on asbestos abatement. Since the late 1980s, a number of studies have contended that asbestos risks in buildings, homes, and schools are not as great as initially believed and that efforts to remove it may actually increase the dangers.

A 1989 report by the Harvard Energy and Environmental Policy Center, for instance, concluded that concentrations of asbestos fibers in schools, and commercial and public buildings are *thousands* of times lower than in occupational settings, where asbestos has been associated with lung cancer, asbestosis, mesothelioma, and other diseases. The report contended that an unwarranted "fiber phobia" had set in among the general public, despite risk assessments showing indoor exposures to ETS and radon are "probably 200 to 400 times larger than even the most conservative (pessimistic) asbestos risk estimates." What's more, the Harvard report suggested that in-place management of intact asbestos in commercial and public buildings may be safer than removing it, which is far more costly and can cause more harm than good by stirring up fibers.

In 1991, a Health Effects Institute–Asbestos Research study commissioned by the EPA echoed the Harvard report, concluding that the presence of asbestos in well-maintained buildings is not a cause for concern. "Asbestos-containing material within buildings in good repair," it found, "is unlikely to expose office workers and other general building occupants to airborne asbestos fiber concentrations above the levels found in air outside such buildings."

Today, most experts agree that asbestos poses virtually no hazard as long as it is in good condition, not damaged or crumbling, and is undisturbed and well maintained. Consequently, state and federal laws have increasingly lent flexibility to licensed abatement contractors to decide whether asbestos can be better handled by encapsulation or enclosure and maintenance, instead of more costly removal. Encapsulation typically involves "painting" a binding or sealing agent onto asbestos to contain fibers. Enclosure involves the wrapping or covering of asbestos, or sealing it behind air-tight barriers. While these techniques are promising in some cases, in certain situations—where asbestos can be damaged by water or is subject to disruption—it is best removed by qualified experts using specialized equipment. In addition, some cases involving real estate transactions or liability insurance may require complete removal of asbestos.

Sources and Reduction Strategies

A complete listing of the many thousands of products that contain asbestos would fill volumes. Consequently, EPA officials and other experts advise hiring licensed professionals—listed with state and local health departments and regional EPA offices—to evaluate any suspected asbestos problems in homes, schools, offices, and other buildings.

Asbestos abatement contractors, who comprise a $3 billion-a-year industry employing some 40,000 workers in the United States, are certified by state and federal officials and use a variety of sanctioned techniques for reducing asbestos risks. In general, asbestos that is in good repair can be sealed or covered. Encapsulation and enclosure is less expensive than asbestos removal, which typically costs $25 per square foot. Asbestos that is damaged, crumbling, or otherwise deteriorating can be safely removed by professionals who are knowledgeable about the health risks, skilled in abatement techniques, and trained to protect the environment during abatement procedures.

When hiring a professional, experts advise, building managers and homeowners should be wary of "rip and skip" artists who fail to take proper precautions and leave behind clouds of asbestos dust. Reference checks and license verifications can trip up ripoff artists, who sometimes hire immigrant workers who may not know that asbestos removal requires special clothing and equipment.

While the EPA and health experts advise homeowners and building managers *never* to attempt asbestos removal projects on their own, they have identified a number of common household products and locations that may contain asbestos. Among them:

• Wall and ceiling insulation installed in buildings constructed between 1930 and 1950, as well as sprayed-on soundproofing and decorative material

• Insulation—including asbestos paper tape and blanket wrapping—on steam heating and hot water pipes; boilers; oil-, coal-, and wood-burning stoves; and furnace equipment and ducts; especially in structures built between 1920 and 1972

• Older vinyl floor tiles, sheet flooring, and ceiling tiles

• Drywall joint finishing, patching compounds, and textured paints manufactured before 1977

• Exterior wall siding, roofing felt, and shingles

• Fire retardant material sprayed onto structural steel building beams

• Some older household appliances, including toasters, ovens, broilers, refrigerators and pre-1980 hair dryers

• Brake linings, chemical filters, and reinforcing agents in cement, vinyl products, and asphalt.

Biological Agents

In 1976, more than two dozen people attending an American Legion convention in Philadelphia died from a severe form of pneumonia that was later traced to a bacterium, *Legionella pneumophila*, spread by the facility's air-handling system. Today, most people are familiar with what has become known as Legionnaire's disease. But what is less well known is that the bacterium, which breeds in building ventilation and plumbing systems, continues to kill thousands of people each year.

Legionella pneumophila, which is fatal in 15 to 20 percent of the people it infects, is one of the most dangerous members in the family of indoor air contaminants known as biological agents, or "bioaerosols." Unlike other indoor contaminants, biological agents are living organisms or the emanations of living organisms. They include such things as microscopic bacteria and viruses, which cause the influenza and the common-cold outbreaks that strike tens of millions of people each year, as well as smallpox, measles, chicken pox, staph infections, and rubella. They include plant-like fungi (mold and mold spores), algae, mildew and pollens, which turn the world into a fuzzy-itchy nightmare for forty million allergy sufferers. And they can even include common components of simple house dust—such as pet dander, dust mite droppings, body parts, and excreta of insects and animals—which trigger hundreds of thousands of emergency-room visits each year by asthmatics.

In the late 1980s, NIOSH ranked biological agents third on its list of leading indoor air pollution problems, behind poor ventilation and building fabric contaminants. In terms of health impacts, however, bioaerosols are the biggest and most pervasive indoor air quality problem facing large buildings and single-family homes.

"From a public health perspective, biological agents are probably the most dangerous of all the indoor air pollutants," observed Dr. Harriet A. Burge, an associate professor at Harvard, in testimony to Congress in 1991. "Influenza alone affects 100 million people each year and probably kills at least 10,000, causing economic losses in excess of $5 billion. Legionnaire's disease probably kills more than 35,000 people a year—and that's a conservative estimate."

"More than 400,000 hospital emergency room admissions each year result from asthma, many of which are related to exposure to biological agents. For 25 to 30 percent of the population, the quality of life is adversely affected by airborne allergens."

Biological agents are common in outdoor air as well as indoor environments but have received little attention because they cannot be controlled the same way regulators can restrict toxic industrial emissions. Yet bioaerosols, while dangerous, are among the most easily controlled contaminants in homes and large buildings through adequate ventilation, HVAC maintenance, and humidity controls.

Sources and Symptoms of Bioaerosols

One of the interesting features about bacterial and viral contaminants is that they are most often brought into the building by the occupants themselves and transported through breathing, coughing, or sneezing. Plants, insects, and pets can also be sources of pollen, dander, and other allergens. Molds and bacteria can grow and flourish in poorly maintained air ducts, humidifiers, air conditioners, dehumidifiers, and even air-cleaning filters. In many cases, poor ventilation and high humidity work in tandem to create or exacerbate bioaerosol problems indoors.

The most common health effects brought on by exposure to biological contaminants are allergic reactions—skin rashes, itching, sneezing, bronchial asthma, hay fever (allergic rhinitis), headaches, malaise, hypersensitivity pneumonitis, and humidifier fever. Infectious diseases—influenza, colds, small pox, tuberculosis, and the like—are also typical in buildings with too little fresh air and too much water vapor, which allow viruses and bacteria spread through the air to invade and grow in human tissues. And in some cases, certain kinds of molds can also produce toxins that are potent carcinogens.

73

Even the cleanest home or building is subject to the spread of biological contaminants, which require very little to survive. For instance, dust mites, microscopic relatives of spiders whose droppings cause allergic reactions in asthmatics, live on human skin scales and thrive in moist environments. Consequently, they commonly inhabit household carpets, mattresses, pillows, linens, blankets, curtains, upholstered furniture, and even stuffed toy animals. Tens of thousands of varieties of mold also flourish in warm, humid environments. Living on plant and animal matter, they reproduce by releasing spores into the air, which can cause allergic symptoms and induce hypersensitivity even in healthy individuals.

Reducing Risks from Bioaerosols

Currently, no regulations or guidelines exist for biological contaminants because researchers have not yet been able to establish definitive "safe" levels for fungal spores, dust mite droppings, and other bioaerosols indoors. But EPA officials, allergists, and IAQ specialists have outlined a series of prudent steps homeowners and building managers can take to reduce the risks of bioaerosol contaminants by maintaining low levels of relative humidity and proper ventilation.

In All Buildings

• Observe ASHRAE standards for adequate ventilation to dilute concentrations of bioaerosols and other contaminants. Maintain clean, well-run air handling units, ductwork, and HVAC systems to ensure proper volumes of fresh air.

• Keep relative humidity levels indoors low through the use of dehumidifiers, air conditioners, and exhaust fans to inhibit the growth of molds, mildews, dust mites, and other microbes. The EPA suggests a range of 45 to 50 percent humidity; ASHRAE recommends levels no higher than 60 percent. Bear in mind that low levels of humidity can leave the respiratory system open to infection.

• Clean and maintain areas where water collects, including drip pans of humidifiers and refrigerators, cooling towers, bathroom and kitchen areas, basements, carpets, and fabrics.

• Exterminate insect populations.

• Routinely clean and replace air filters on furnace, air conditioning, and air cleaning devices.

In the Home

• Control dust mites by encasing mattresses, box springs, and pillows in zippered, dust-proof covers, and wash pillows and blankets in hot water every two weeks. Carpeting and curtains, common breeding grounds for mites, can also be removed or treated with substances that "denature" allergens.

• Install exhaust fans in kitchens and bathrooms to remove water vapors and inhibit the growth of biological agents; the average family of four generates up to ten gallons of water vapor a day simply through breathing, sweating, cooking, showering, and other household activities. For increased ventilation throughout the home, consider installing a whole-house fan.

• Observe good housekeeping. Frequently clean kitchens and bathrooms with mold-killing solutions—a mixture of equal parts bleach and water, for instance—to curtail microbial growth. Vacuum carpets regularly, particularly in bedrooms, where mite growth is highest. High-efficiency particle air filters and vacuums can minimize the presence of bioaerosols indoors.

• Clean and disinfect free-standing cool-mist and ultrasonic humidifiers—common breeding grounds for biological agents—on a daily basis and use only distilled water. Humidifiers in centrally housed units should be checked and cleaned frequently.

• Dehumidifiers and air conditioners, when serviced and cleaned regularly, can keep humidity levels low indoors.

• Vent clothes dryers outdoors and ventilate attics and crawl-spaces, which can provide ideal conditions for mold.

• Do not store clothing or other items likely to harbor mold in damp basements or attics.

• Be particularly vigilant in areas that can encourage mold growth, such as greenhouses, saunas, and summer cottages.

• Repair roof leaks promptly and ensure proper water drainage to avoid pooling at the foundation of the home.

CHAPTER 3

Living With Chemicals

Volatile Organic Compounds

In 1987, nearly a decade after events at Love Canal raised profound new questions about the public health risks posed by toxic chemicals, the EPA released what would become a landmark study of personal exposure to some of the most common contaminants at Superfund sites across the country. Based on a series of surveys begun in 1979, the EPA's Total Exposure Assessment Methodology (TEAM) study was designed to document and characterize actual human exposures to the family of substances known as volatile organic compounds (VOCs). But what EPA researchers found raised a number of unexpected questions about federal air pollution policy that were at least as profound and troubling as those raised by Lois Gibbs and other Love Canal protesters ten years earlier.

Headed by EPA scientist Lance Wallace, the main TEAM study involved six hundred participants in four states—New

77

Jersey, North Carolina, North Dakota, and California—chosen to represent a total population of 717,000. Each participant was asked to carry a personal air sampler through a typical twenty-four-hour day to sample the air breathed at work and at home, indoors and out. Separate samplers measured outdoor ambient air quality down to parts-per-billion levels, and each participant also contributed samples of exhaled breath. Although the findings of the study came as no surprise to veteran IAQ experts, they were something of a shock to EPA insiders and policy wonks. Among them:

• Mean personal exposures to eleven target VOCs examined were greater indoors than outdoors at seven of eight monitoring sites.

• Indoor concentrations were typically two to five times greater, with some up to one hundred times higher—regardless of whether participants lived near industrial facilities or in rural, residential areas.

• VOC sources were not neighboring toxic waste sites or factories but common household items such as cigarette smoke (benzene, xylenes, ethylbenzene and styrene); tap water (chloroform); air fresheners, toilet bowl deodorizers, moth crystals, air fresheners (paradichlorobenzene); dry-cleaned clothing (tetrachloroethylene); and gasoline and auto exhaust (benzene). Other sources identified by researchers included furniture, solvents, paints, drapes, cosmetics, construction materials, wallpaper, linoleum, paints, adhesives, carpeting, and pesticides.

• For all but one of the chemicals, inhalation accounted for 99 percent of the participants' exposures.

Other Evidence

The EPA's 1987 report was perhaps the most widely publicized of its indoor air studies targeting VOCs. But it was by no means the agency's final word on indoor exposures to chemicals.

• In 1988, an EPA survey of the air quality inside ten public buildings turned up traces of more than five hundred chemicals —two dozen of which are known or suspected carcinogens and some of which reached concentrations one hundred times typical outdoor levels.

• In 1989, an EPA report to Congress on indoor air expanded on the issue of indoor VOCs and raised questions about the synergistic effects of a *combination* of contaminants from building materials, consumer products, furnishings, pesticides, and fuels. "More than nine hundred different volatile organic compounds have been identified in indoor air," the report stated. "Exposure to mixtures of VOCs commonly found in building materials may be an important source of sick building syndrome complaints."

• In 1990, EPA researchers found that many common indoor activities—washing dishes, using bathroom deodorizers, cleaning an automobile carburetor—can result in ten- to thousand-fold increases in eight-hour exposures to specific VOCs.[8]

What all of these studies demonstrated is that many of the same chemicals that have raised concerns among environmentalists and regulators when found at the nation's 1,200-plus Superfund sites are common in virtually all indoor environments at levels far greater than those typically found outdoors.

Summarizing the implications in a 1990 report to Congress, EPA officials observed: "A growing body of scientific evidence indicates that the air within homes and other buildings can be more seriously polluted than outdoor air, even in the largest and most industrialized cities."

Sources and Health Effects of VOCs

VOCs are a class of carbon-based chemicals that volatilize, or evaporate, at room temperature. Early studies of VOCs indoors focused on building materials and adhesives in the 1970s, then expanded to include household cleaners, consumer goods, and other sources. Formaldehyde is the most common VOC in indoor air, but a handful of others are also typically found in homes, offices, schools, and buildings. The most common sources of VOCs are:

Solvents—including lacquers, cleaners, adhesives, paints, strippers, degreasers, and cosmetics, which can contain benzene, toluene, methyl and ethyl compounds, mineral spirits, and ethers.

Organochlorines—including pesticides, cleaning agents, and preservatives such as 1,1,1-trichloroethane, trichloroethane, carbon tetrachloride, chlorobenzene, heptachlor, and chlordane, pentachlorophenol, and hexachlorophene. PCBs and vinyl chloride are also in this class.

Phenols—including coal tar and petroleum compounds, disinfectants, antiseptics, plastics, cleaners, perfumes, air fresheners, mouthwashes, polishes, waxes, and glues.

Occupational studies and animal tests have shown that exposure to some VOCs can cause both acute and chronic health effects. Symptoms of VOC exposure can include fatigue, headaches, dizziness, weakness, joint pain, numbness in extremities, euphoria, blurred vision, and irritation of the eyes, nose, throat, and skin. Many VOCs are asphyxiants, some are potent narcotics, and a handful are neurotoxic. At high concentrations, some have been shown to cause liver and kidney damage. A few VOCs are known or suspected carcinogens, with the EPA estimating that just six common VOCs contribute to one thousand to five thousand excess cancer cases per year nationwide.

VOCs have also been implicated in "multiple chemical sensitivity" disorders, where individuals have allergic-like reactions to certain chemicals after a single sensitizing dose or a sequence of chronic exposures. "Individuals who appear to demonstrate multiple chemical sensitivity report severe reactions to a variety of VOCs and other organic compounds, which are released by building materials and various consumer products including cosmetics, soaps, perfumes, tobacco, plastics, dyes, and other products," EPA researchers stated in a 1991 reference manual on indoor air. "Many of the chemicals contained in these products are potent sensitizers."

VOC exposures can be particularly acute during painting and cleaning activities, when concentrations can reach 100 ppm in poorly ventilated indoor settings. In addition, some VOCs can be absorbed by building materials and fabrics and reemitted over a period of years, posing potential longterm exposure risks. Currently, the role of VOCs in sick building syndrome is under debate. But EPA and NIOSH investigators believe VOCs may play a leading role in many cases.

EPA researchers have singled out several VOCs as the most important and prevalent in indoor air:

• *Benzene,* a respiratory irritant and human carcinogen, is present in plastic and rubber solvents, paints and varnishes, cigarette smoke, and gasoline.

• *Methylene chloride* is listed by the EPA as a probable human carcinogen which, once inhaled, is converted to carbon monoxide in the blood. It is commonly used as a paint stripper but is also present in more than six hundred consumer products tested by the EPA.

• *Aromatic hydrocarbons*—including xylenes, toluene, styrene, ethylbenzene, naphthalene—are neurotoxic at high concentrations and can cause irritation, headaches, fatigue, and stupor. Products that typically include such compounds are paints, adhesives, solvents, and fuels.

• *Halogenated hydrocarbons*—such as tetrachloroethylene, trichloroethylene, 1,1,1-trichloroethane, dichlorobenzenes, and chloroform—include a number of known or suspected carcinogens and are present in cleaning solvents, waxes, and degreasers.

Risk Reduction Strategies

As with other indoor contaminants, increased ventilation and source controls are twin strategies recommended for reducing risks from VOCs indoors. New homes and buildings, as well as those undergoing renovations and paint jobs, should be ventilated at higher levels until VOC-containing products and materials dry and cure. In all cases, products containing VOCs should only be used in accordance with manufacturers' specifications.

Careful selection of cleaning products, as well as building materials and furnishings, can also reduce VOC levels indoors. In fact, an entire "green" industry has sprouted up in recent years marketing non-toxic cleaners. Greenway Products Inc. and Ecover Inc. both offer full lines of such products; others are available by mail order through Seventh Generation, a Burlington, Vermont–based company. The American Lung Association has also produced a consumers guide that provides homemade "recipes" for non-toxic alternatives to chemically based consumer goods. Several books are also available detailing non-toxic products and alternatives to commercially available chemical products.

In addition, the EPA and other health experts advise:

• Limit the use of VOC-laden products indoors, including chemical air fresheners, deodorizers, solvents, cleaning products, sprays, and other items.

• Store paints, pesticides, fuels, and other toxic products outside the building or in well-ventilated areas. Buy only what is needed for a particular project and throw away only partially full containers.

• Restrict certain activities—such as smoking, painting, furniture refinishing, and other hobbies involving products that emit VOCs—to well-ventilated areas.

• Plan paint jobs, pesticide applications, and other applications of VOC-laced products when homes and buildings are unoccupied, such as during the evening or on weekends, when possible. Also schedule renovations during off-hours and increase building ventilation—over-ventilate—during and after such projects. Some building experts also advise "baking out" new and renovated buildings by turning up the heat to 90 degrees or more to accelerate emission of VOCs.

• Read product labels for contents and choose less- and non-toxic alternatives such as latex and water-based paints over oil paints. Avoid oven cleaners made with lye, air fresheners made with phenols (which kill odors by dulling the sense of smell), and mothballs made with paradichlorobenzene.

• Air out freshly dry-cleaned clothes, which are often laced with perchloroethylene, in ventilated areas for several days.

• Ask retailers for information on VOC emissions from building products, furnishings, carpeting, and other materials before making purchases. Many manufacturers, including the carpet industry, have voluntarily moved to produce products with low-emission rates in the wake of growing IAQ concerns.

Formaldehyde

Few indoor contaminants have generated as much concern and controversy as formaldehyde. The most common volatile organic chemical found in indoor air, formaldehyde has been the subject of a bitter public health debate since the late 1970s, when the Chemical Industry Institute of Toxicology linked it to cancer in rats. Although the EPA has labeled formaldehyde a probable human carcinogen, health studies have produced conflicting conclusions about its carcinogenicity in humans and left the scientific community largely divided over its cancer risks.

What is not in dispute is that formaldehyde is a potent mucous membrane irritant that can cause eye, nose, and throat discomfort, coughing, dizziness, headaches, and skin rashes. Most people can't detect its presence below concentrations of 1.0 part per million. But some researchers have found sensitive persons can identify the substance at levels as low as 0.05 parts per million. Because it is used in a plethora of industrial and manufacturing processes, it is ubiquitous in indoor environments.

A colorless gas with a pungent odor, formaldehyde was first identified as a potential "bad actor" in indoor air in 1982 when the Consumer Product Safety Commission tried to ban the use of urea formaldehyde foam insulation after it had been sprayed into the walls of between half a million and one million American homes. Until that time, formaldehyde had been largely hailed as a miracle industrial chemical, inexpensively synthesized in the laboratory.

Formaldehyde was first used commercially as a preservative and disinfectant in hospitals and labs in the Nineteenth Century. Today, more than six billion pounds of it are produced in the United States each year for use in everything from glues and fuels to textiles and paper. About half the formaldehyde produced is used in resins of bonding and adhesive agents in building materials and wood products, such as paneling, plywood, particleboard, fiberboard, wallboard, cabinets, furniture, hardboard partitions, and ceiling panels. An estimated three billion square feet of particleboard are produced each year, two billion square feet of hardwood plywood, and 600 million square feet of medium-density fiberboard.

Other common products that contain formaldehyde include:

• Clothing, linens, draperies, fabrics, upholstery, and other materials produced by the textile industry that are crease-resistent, fire-retardant, color-fixed, and waterproof.

• Pre-pasted wallpapers, fiberglass insulation, and ceiling and floor coverings.

• Paper products—including facial tissues, napkins, waxed paper, grocery bags, and paper towels—that derive their strength from the chemical.

• Personal care items, consumer goods, and household products such as soaps, shampoos, cosmetics, antihistamines, antiperspirants and deodorants, pharmaceuticals, toothpaste, diaper liners, dyes, insecticides, paint, air fresheners, glues, plastics, furnishings, carpets, waxes, oils, and cigarette smoke.

Health Effects

In small quantities, formaldehyde is not believed to be toxic.

But in closed environments, it can build up and be inhaled, ingested, and absorbed by the skin. Although the primary health effects of exposure to formaldehyde are eye, nose, and throat irritation, it has also been shown to produce wheezing, coughing, diarrhea, nausea, headaches, lethargy, dizziness, sleep disorders, nosebleeds, vomiting, and rashes, and to trigger asthma. Some controversial studies have linked it to menstrual irregularities, neuropsychological problems, and respiratory disease.[7]

The EPA has found that most people become irritated by formaldehyde at levels between 0.1 and 1.1 ppm, but a National Academy of Sciences committee determined 10 to 20 percent of the population—including infants, seniors and other vulnerable persons—may respond at lower levels of exposure. Several studies have also suggested exposure to formaldehyde may cause some individuals to become "sensitized," such that they subsequently react to progressively lower exposure levels.

Reducing Exposure to Formaldehyde

Indoor exposure to formaldehyde is determined by two key factors: the rate at which it is emitted from products and the rate at which it is removed through ventilation. Heat and humidity can increase the rate at which formaldehyde "off-gasses" from products, a process that can occur over months or years after a product is manufactured. Typically, emissions decrease as materials cure and age, with the bulk emitted the first six months to a year after manufacture. For this reason, new energy-efficient homes built with bonded wood products are more likely to have higher levels of formaldehyde than older, conventionally constructed houses. Studies by Lawrence Berkeley Laboratories and the Consumer Product Safety Commission (CPSC) have found the average formaldehyde level in a conventional home to be about 0.05 ppm, while newer ones average 0.1–0.2 ppm.

Formaldehyde exposures can be reduced by increasing ventilation, choosing materials made with little or no formaldehyde, and applying sealants and encapsulants to cabinets, particle-board, and other products. Among the specific risk-reduction strategies the EPA and others advise:

• Increase ventilation in buildings that have undergone recent renovations or have been newly furnished, carpeted, or decorated. Buildings at special risk include the nation's nine million mobile homes, which are chiefly constructed of bonded wood products, as well as newly refurbished homes, offices, schools, and other buildings with mechanical ventilation systems.

• Buy and use products—including building materials and furnishings—that emit low levels of formaldehyde. Information about product emissions levels is usually available from manufacturers. In general, exterior-grade bonded wood products contain glues that have lower emission rates, as do solid wood items. Under regulations promulgated by the U.S. Department of Housing and Urban Development (HUD) in 1984, formaldehyde emissions from bonded wood products used in mobile homes are limited to between 0.2 ppm and 0.3 ppm. Pressed wood and particle board that meet the HUD standard will be so labeled.

• Emission barriers such as sealing paints, polyurethane coatings, veneers, and decorative overlays can be used to limit formaldehyde emissions from pressed wood products.

• Combustion sources that emit formaldehyde—gas stove appliances, for example—can be addressed through the use of exhaust fans in kitchen areas. Restricting cigarette smoking, which also produces the chemical, can also limit levels.

• Chamber tests conducted by NASA have found more than a dozen common houseplants, including the elephant ear philodendron, spider plant and golden pothos, have some limited ability to filter out formaldehyde in indoor settings.

• To determine indoor formaldehyde levels, inexpensive monitors can be purchased (usually for under $50) and installed.

Chapter 3

Pesticides

In 1964, Rachel Carson's *Silent Spring* launched the modern environmental movement with a compelling indictment of the use of pesticides, particularly DDT, in the United States. Yet today, while DDT has long been banned, the U.S. pesticide industry manufactures at least twice as many pesticides it did in the early 1960s, reporting more than $6 billion in annual sales. What's more, nonagricultural uses of pesticides—accounting for a quarter of the nation's consumption, according to EPA estimates—are rapidly increasing, with some surveys estimating chemical lawn service sales have nearly tripled since the 1970s.

Chemical pesticides are certainly beneficial in many applications and have become a necessary staple in the nation's efforts to promote good health, control disease, and feed its people. But chemical pesticides also have several characteristics that make them a human health risk. First, they are by design toxic to living things—including insects, rodents, fungi, and weeds—and can therefore pose a threat to human populations as well. Secondly, many persist in the environment for years, as evidenced by the presence of chlordane, a termiticide banned by the EPA in the late 1980s, in many of the estimated thirty million American homes treated with it. Finally, pesticides are fat soluble and can accumulate in the fatty tissues of living things.

Regulation of Pesticides

Pesticides are classified as insecticides, herbicides, fungicides, nematocides, plant growth regulators and biological control substances. About 60 percent of the market is insecticides, 35 percent is herbicides, and the rest comprise the remaining 5 percent. Agricultural uses are the biggest, accounting for 77 percent (one billion pounds a year) in the United States, according to EPA estimates in the late 1980s. The remainder include applications by utilities, lawn services, homeowners, gardeners, restaurants, stores, golf courses, offices, and buildings.

Pesticides are primarily regulated under the Federal Insecticide, Fungicide, and Rodenticide Act of 1947 (FIFRA), overseen by the EPA, in cooperation with OSHA, the Department of Agriculture, and the Food and Drug Administration. Regulated pesticides include everything from farm chemicals to disinfectants that kill bacteria in homes and hospitals, to food additives, to chemicals that retard the growth of bacteria, fungi, and other living organisms in leather, cloth, and wood products.

Ingredients in pesticides are generally classed as active (toxic) or inert (fillers, binders, enhancers). But in 1987, the EPA identified 57 inerts which have known or suspected toxic effects. Most of the more than 20,000 registered pesticides on the market are formed from combinations of 1,800 active ingredients, which are themselves grouped into about 600 classes. Unfortunately, most of these substances were approved for use prior to the EPA's adoption of strict registration regulations that now require all new pesticides to undergo rigorous toxicity testing. According to the National Academy of Sciences, only 10 percent of pesticides in use had undergone complete hazard assessments as of 1984.

Under amendments to FIFRA approved in 1988, the EPA is required to re-register about 1,400 of those active ingredients by 1997 for health and environmental assessments using current criteria. But because of delays in implementation, environmentalists charge that many pesticides with potentially serious health risks may remain on the market for years before coming under EPA scrutiny.

To date, the EPA has identified about sixty pesticide chemicals as possible carcinogens. It has also banned or restricted a handful of "bad actors"—including DDT, chlordane, aldrin, dieldrin, lindane, heptachlor, and 2,4,5-T—and barred the indoor use of the wood preservatives pentachlorophenol and creosote.

Indoor Exposures

Although media reports and regulatory efforts have tended to focus on the risks of pesticides in agricultural settings and in foods, mounting scientific evidence suggests many people also face significant exposures in indoor environments. For instance:

• In 1988, the EPA's Non-Occupational Pesticides Exposure Study (NOPES) of the air inside more than two hundred homes in two Massachusetts and Florida cities found traces of eleven pesticides. The most common were chlorpyrifos (Dursban), chlordane, heptachlor, diazinon, and propoxur. Echoing earlier EPA TEAM studies, researchers also found "mean outdoor air concentrations were almost always lower than mean indoor and personal concentrations."

• In 1991, an EPA report to the American Cancer Society suggested house dust may be a "major route" of infant and toddler exposures to pesticides—including many of those cited in the 1988 NOPES study. The study, which detected residues of thirty common pesticides in the house dust and yard soil of nine middle-income homes, found not only traces of chemicals that had been recently applied, but also residues of pesticides that had not been used on the premises for some time. What's more, researchers noted "levels of many targeted pesticides in soil outside the home were lower than those found in house dust." This, they suggested, indicated that "while lawn and garden chemicals may have relatively short lifetimes in the outdoor environment, residues tracked into the home would be assumed to remain unaltered for years."[9]

In addition, the EPA has reported:

• Nine out of ten homes have at least one pesticide in or around the house—in the form of pest strips, baits, dusts, "bug bombs," powders, aerosol sprays, solutions, and skin repellents—but national surveys show only 50 percent of people who use them read labels for application procedures.

• Eleven percent of single-family households use

commercial lawn care services, with the Government Accounting Office estimating lawn care pesticides account for sixty-seven million pounds of active ingredients used annually—8 percent of those applied in agricultural settings.

• Home and garden applications account for almost 29 percent of the 230 million pounds of pesticides used each year for nonagricultural purposes; the remainder are used by industry, government, and commerce, according to the GAO.

• More than 57,000 cases of pesticide exposure were reported to poison control centers in 1987—98 percent of them due to accidental exposures, and 60 percent involving children under six. From 1980 to 1985, half of all pesticide-related deaths occurred in the home, making pesticides the second most common cause of childhood poisonings.

• Seventeen different pesticides—including potential human carcinogens aldicarb and ethylene dibromide—have been found in the groundwater of twenty-three states.

Health Effects

Chemical pesticides have a wide range of toxicities and potencies and, at high levels, can cause neurotoxicity, teratogenicity, liver damage, cancer, and other health effects. Pesticides can be inhaled, ingested, or absorbed through the skin. In determining a pesticide's toxicity, manufacturers must identify dosages needed to kill half of laboratory test animal populations via all three routes to establish what is called an LD (lethal dose) 50. The lower the LD 50 value, the more potent the pesticide. In addition, researchers use animal tests to determine other adverse effects, such as their ability to cause nerve damage, cancer, gene mutations, birth defects, and tumors.

Common effects of pesticide exposure include eye, nose, and throat irritation, rashes, and flu-like symptoms. But in high and chronic doses, some pesticide chemicals have been linked to neurological effects, liver and kidney damage, gastrointestinal problems, birth defects, and cancer. Some pesticides, including organophosphates and organochlorines, have also been shown to affect more than one organ system.

The EPA requires pesticides to be labelled according to one of three toxicity categories: *danger* (highly toxic), *warning* (moderately poisonous), and *caution* (least hazardous). The labels, however, provide only minimal information and do not suggest the degree of chronic health risks associated with specific products. Some states have moved to toughen regulations on pesticides, with many states requiring the posting of warnings for the application of lawn chemicals and other pesticide uses. But even in states with the most aggressive laws, many aspects of pesticide handling and exposure in nonagricultural settings depend on the activities of tenants, homeowners, and building managers. And, as EPA surveys have shown, consumers tend to be casual about pesticide use, perhaps because so many do not require a license to apply and are readily available in grocery stores and hardware stores.

Reducing Pesticide Exposure Risks

The Department of Food and Agriculture has estimated that nearly one-fifth of all pesticides on the market are sold to homeowners, a class that is "undoubtedly the least knowledgeable in the proper use and handling of pesticides." Consequently, the best way for most Americans to protect themselves from pesticide exposures is to arm themselves with information. That means reading labels and asking questions of pesticide applicators and local, state, and federal health officials. The same rules apply for managers of buildings and grounds where pesticides are used.

The EPA and the Department of Food and Agriculture advise a number of specific recommendations to control indoor pesticide exposures. Among them:

* Use non-toxic alternatives to chemical pesticides where

possible, and compare all options before making a decision. For instance, lawn and gardening services can often use "Integrated Pest Management" methods, which minimize the use of chemicals and utilize "good bugs" and pathogenic substances as alternatives.

• Choose pest-resistent plant varieties in landscaping and observe good horticultural practices, such as proper mowing and fertilizing, to keep lawns thick and healthy.

• Consider using organic fertilizers—derived from chickens, cows, fish, and humans—in place of chemicals.

• Read labels of pesticide products and heed manufacturers' instructions on the use of protective clothing, ventilation during application indoors, measurement, usage, and storage.

• Never eat, drink, or smoke when handling pesticides.

• Buy only as much of a product as is needed and store all unused portions away from children. Where possible, keep chemical products in outdoor or well-ventilated areas such as garages.

• Reduce the need for pesticides by filling holes and cracks in walls, particularly in semi-outdoor areas such as garages, and storing wood scraps away from buildings. Also maintain good housekeeping; don't leave food and dirty dishes lying around.

• Keep structures dry to reduce wood rot, which can attract ants and other pests.

• When using a professional pesticide service or lawn care company, check references and ask for detailed informational brochures on the chemicals used. Calls to state health officials and the EPA can provide details about the risks of chemicals.

• When spraying outdoors, keep children and pets away— usually for forty-eight hours—and remove toys and lawn furniture.

• For emergency use, keep the number of the local Poison Control Center in a handy location. For more information, the EPA has also established a twenty-four-hour hotline: 1-800-858-PEST.

CHAPTER 4

Home, Healthy Home: A Guided Tour

The sanctity of the home is a concept that transcends time, culture, and religion. Literature tells us home is sweet, it is where the heart is, and—as a famous Kansan once said—there's just no place like it. Yet scientific researchers are becoming increasingly convinced that the modern American home is not the safe, nurturing haven from a hostile world long depicted in historical, literary, and religious texts. In fact, when it comes to our health, the typical home of the 1990s may be more aptly characterized by Pogo: "We have met the enemy and he is us."

Indeed. Radon, which kills fourteen thousand Americans each year, contaminates the air in one of every fifteen American homes. Volatile organic compounds, which cause up to five thousand cancers annually, are common in household products and building materials. Lead-based paints poison one in six

American preschoolers in their homes. And that's not to mention carbon monoxide, asbestos, secondhand smoke, pesticides, biological agents, and other indoor contaminants that collectively account for thousands of deaths and millions of illnesses a year.

This, obviously, is the bad news. Here's the good news:

When it comes to controlling exposures to toxic substances and air pollution, Dorothy had it right—there *is* no place like home. In fact, most people have far more power to limit their exposure to toxic contaminants at home than anywhere else—outdoors, at work, or in transit. It's often just a question of knowing what to look for, where to turn for answers on indoor air pollution, and how to maintain safe, healthful conditions in every room in the house—from the ground up.

The Basement

Few places in the home are likely to have as many different potential indoor contaminant sources and problems as the typical basement. The lowest area in the home, the basement can harbor radon from soil gases infiltrating through cracks, drains, gaps, and other openings in walls and floors. Faulty heating units can lead to oil and natural gas leaks or spew carbon monoxide and other combustion gases into the air. Mold, insects, and other biological agents can thrive in damp, dark basement areas. Sewer gases can be released from damaged or aging plumbing systems, sumps, and floor drains. And stored fuels, paints, cleaners, pesticides, and other chemical products produce harmful vapors that can fill basement areas or be sucked up into the living spaces of the home because of air pressure factors.

Fortunately, several relatively simple steps can be taken to limit the buildup of chemical, biological, and naturally occurring gases, particles, and contaminants in basement areas by controlling pollution sources and increasing ventilation.

• Radon tests can determine whether a home has levels of

the naturally occurring cancer-causing gas that are higher than the EPA's safe cutoff of 4.0 picocuries per liter (pCi/L), which poses the equivalent risk of two hundred chest X-rays per year. A short-term test should be conducted, preferably in winter when radon levels are highest, in the lowest living space in the home (the basement if it is used frequently; otherwise the first floor). If test results are above 4.0 pCi/L, but lower than 20.0, homes should be retested for verification and fixed within a year. Immediate action should be taken in homes with levels higher than 20.0 pCi/L, which pose lung cancer risks greater than those faced by a pack-a-day smoker. Homes can be fixed by sealing cracks and boosting ventilation through "sub-slab depressurization."

• Leaks of gas and combustion gases are common and not always easily detected, from natural gas water heaters, furnaces, clothes dryers, and cracked or blocked exhaust flues. Annual checks, usually done free of charge by utilities, can identify and ward off problems. Leaks can also be determined by applying soap-and-water solutions to valves, joints, and control panels; bubbling indicates the presence of a leak. Be aware of the warning signs: chronic cold- and flu-like symptoms, headaches, and disorientation can be symptoms of carbon monoxide poisoning. Irritation of the eyes and respiratory tissues can indicate a problem with particulates from combustion appliances, including wood-burning units, that are improperly vented indoors.

• Oil-burning units can also leak, creating fumes or pools of fuel in basement areas that produce volatile chemical vapors. Annual checkups by home heating oil companies and plumbers can identify problems before they escalate in heating units and plumbing. Spilled oil should be cleaned with detergent and/or sealed with floor paints.

• Keeping basements dry and clean can deter insects, mold, and biological growth. To limit moisture problems indoors, be sure there is proper water drainage from gutters and downspouts, so there is no pooling around a home's foundation. Use a dehumidifier to keep humidity levels down in damp basements. Carpeting should be avoided in basements that are prone to flooding and moisture problems. Cracks and leaks that allow water to enter should be sealed. Walls and floors can be waterproofed. Moldy areas can be cleaned with a weak water-chlorine solution; non-toxic cleaners, such as borax, can inhibit biological growth.

• Chemical products typically stored in basement areas—including pesticides, fuels, cleaners, paints, strippers—can release vapors from volatile organic compounds. It's a good idea to store such products in outdoor areas where possible or in cabinets that are ventilated to the outside.

• Pesticides applied indoors and to foundations of homes outdoors can persist for years; chlordane, for instance, can still be detected in many of the millions of homes treated with it before it was banned by the EPA in the 1980s. This can create a potential risk to children, especially infants and toddlers, whose hand-to-mouth activities render them especially vulnerable to chemicals in house dust and yard soils. To limit pesticide needs, bar termites and ants by installing metal barriers around home foundations, sealing gaps and holes in walls, and storing wood away from the house. Several emerging alternatives to chemical pesticides also appear to hold promise, including new methods that use heat and electric jolts to kill bugs and "biopesticides," such as the EcoScience Corporation's Bio-Path Cockroach Control Chamber, which uses a natural microbial agent to kill cockroaches.

The Kitchen

Like the basement, the kitchen of a home can harbor a large variety of potential indoor air pollution sources. Gas ranges emit a host of combustion gases and chemical contaminants, including carbon monoxide, nitrogen dioxide, sulfur dioxide, and formaldehyde. Refrigerator drip pans, garbage pails, sinks, and gaps in countertops, walls, and floors can provide ideal breeding grounds for mold, mildew, bacteria, and other biological growths that thrive on moisture and food wastes. Bonded wood and particleboard cabinets can emit high levels of formaldehyde gas, especially in the first six months to a year after installation. Some appliances, including blenders and mixers, can produce small quantities of ozone. Many common cleaning products used in the kitchen contain volatile organic compounds and solvents. And even hot water from taps and dishwashers can release volatilizing chloroform into the air.

In tackling all of these problems, exhaust fans that vent to the outdoors can significantly reduce concentrations of indoor pollutants. Keep in mind, however, that such fans can dramatically change air pressures inside tightly sealed energy-efficient homes, causing air to "backdraft" from chimneys and soil gases (such as radon) to be drawn up from basements. To increase ventilation and avoid such air-pressure changes, open a window when using exhaust fans.

In addition, a number of specific steps can be taken to control individual contaminants at the source.

• Buying kitchen cabinets and furnishings made from solid wood, metal, exterior-grade or low-emitting (HUD certified) pressed wood products can limit levels of offgassing formaldehyde, an irritant that has been shown to cause cancer in rats. Particleboard can also be sealed with plastic and waterproof enamel paints and sealants.

• Using a gas stove with an electronic-spark ignition, common in most newer models, instead of a continuously burning pilot light, can substantially reduce combustion gas emissions and concentrations indoors. Roughly a third of all carbon monoxide and nitrogen dioxide emissions from gas stoves are emitted from pilot lights. Install and use an exhaust fan that vents to the outdoors when using a gas stove and/or open a window. *Never* use the stove to heat a home or apartment.

• Seal wall joints, cracks, and gaps in countertops with

caulking to limit spaces that encourage biological growth. Clean refrigerator drip pans regularly. And keep all kitchen surfaces free of food and wastes, which can attract insects.

• Chemical emissions from plastic floorings and adhesive products can be avoided by installing ceramic tile or hardwood floors. Avoid carpeting in the kitchen, which can get wet and harbor biological agents.

• Don't store chemical cleaners, pesticides, mothballs, air fresheners, or household products under the kitchen sink. Keep them outdoors or in garage areas, where possible, and away from children and pets who are naturally curious and easily poisoned. Limit use of chemical cleaners indoors, especially aerosol products that disperse small respirable particles and chemicals throughout the air. Better yet, use non-toxic alternatives; a 1988 Harvard study found higher levels of paradichlorobenzene, methyl chloroform, tricholoroethylene, and decane in homes with mothballs and air fresheners than in those without.

The Bathroom

The biggest indoor pollution problems facing most household bathrooms are due to high temperature and humidity. All indoor air has some water vapor, with the average household generating up to ten gallons a day simply from breathing, sweating, cooking, cleaning, and bathing. But because the primary activities in bathrooms involve water in some way—showering, bathing, washing, flushing—they tend to have the highest levels of water vapor of any room in the house. Consequently, they can provide warm, moist havens for mold, mildew, viruses, bacteria, dust mites, and other biological agents that produce foul odors and can cause allergic reactions and chronic health problems.

With the exception of some areas of the midwestern United States—where relative humidity can drop below 10 percent, leading to increased respiratory infections and airborne dust—household humidity problems in most regions of the U.S.

generally stem from too much water in the air and too little ventilation.

The EPA and indoor air specialists generally recommend indoor environments be kept between 30 and 60 percent relative humidity, a level that discourages fungal, bacterial, and viral growth. Excess humidity is best handled by increasing ventilation; better insulation can reduce dryness in the Midwest and during winter weather. A number of other specific steps can be taken to keep bathroom humidity at optimal levels:

• Install an overhead exhaust fan that vents to the outside and open a window while showering, an activity that produces huge volumes of water vapor.

• Stem water leaks from tubs, showers, baseboards, sinks, and toilet tanks by caulking seams, joints, and gaps.

• Check mold and mildew growth on bathroom tiles, walls, and ceilings by washing such areas frequently with soapy water, borax, or vinegar. If necessary, solutions of water, chlorine, and TSP can kill and inhibit mold and mildew. Avoid using strong chemical cleaners in enclosed, unventilated bathrooms.

• Frequently clean shower curtains, bath mats, and other removable bathroom fabrics to inhibit microbiological growth.

• When renovating bathrooms, keep in mind that ceramic tile on walls, ceilings, and floors are the least susceptible to water damage and biological growth. Paints that are waterproof and resist mold and mildew can also be used.

• In some homes supplied with water from private deep-rock wells, radon gas can be found in bathrooms because of dispersion through showering. Some chemical contaminants—chloroform, for instance—can also be high in some bathrooms. In both cases, inhalation exposures are more risky than ingestion, and water-filter systems can be installed to reduce hazards.

The Living Room and Other Common Areas

Common areas—living rooms, dining rooms, hallways,

stairways, and the like—are typically home to a variety of household fabrics, furnishings, and products that add to the chemical pollution in a home. Book shelves, wall units, and furniture made from pressed wood products, as well as new building materials themselves, can emit formaldehyde gas. Carpets, backings, and adhesives can load a room with vapors from dozens of volatile organic compounds—especially in the first few months after installation. Dry-cleaned curtains and fabrics release perchloroethylene. Sofas and chairs covered with natural and synthetic fabrics that have been treated with chemicals to resist wrinkles, insects, and fungi can add to the chemical load.

What's more, home fireplaces and wood-burning stoves produce potent carcinogenic particles and combustion gases that can build to dangerous levels in poorly ventilated areas. Even the most immaculately kept homes accumulate dust, which is often laced with pesticides tracked in from outdoors, pet dander, insect parts, dust mites and droppings, and even particles from interior and exterior lead-based paints.

Eliminating the risks posed by these myriad substances would be an impossible task. But a combination of good housekeeping, careful selection of household goods, and adequate ventilation can greatly reduce the risks posed by chronic exposure to these common pollution sources.

• When buying wood-based products—furniture, paneling, flooring, shelving—keep in mind that formaldehyde resins in glues used in bonded-wood materials can emit the volatilizing chemical for months, or even years after installation. Choosing solid wood products, exterior-grade pressed wood, or particleboard that meets HUD requirements for low-emitting materials can keep levels low. Controlling heat and humidity, which increase formaldehyde offgassing when high, can reduce exposure risks. And sealing up cracks, gaps, and holes with spackling compound and sealing paints limits offgassing from pressed wood building materials. Some states provide assistance or conduct formaldehyde tests at little or no cost to residents. Some household indoor air quality inspectors also perform these services.

• When purchasing products containing volatile organic compounds (VOCs)—carpeting, rugs, furniture, fabrics—ask questions of retailers and manufacturers about emissions that can affect indoor air quality. The Carpet and Rug Institute, for instance, has adopted a voluntary policy to produce low-emitting carpet, backing, and adhesive products under its "Green Tag" program.

• When possible, choose products made with fewer chemicals and natural fibers. And when bringing home materials that have been treated with chemicals—such as dry-cleaned clothes, curtains, and other fabrics—air them out in well-ventilated or outdoor areas before placing them in common indoor areas.

• Make sure fireplaces, wood-burning stoves, kerosene heaters, and other sources of combustion gases and particles are in proper working order and used with adequate ventilation. Routine inspections can head off "downdrafting" chimneys and clogged flues. Keep in mind that kerosene heaters are *never* to be used indoors without an open window.

* Limit smoking indoors, particularly around children, who are at greatest risk from the health impacts of secondhand smoke. Tobacco smoke is also believed to increase the risks posed by radon. Ask smokers to puff outdoors or on the porch, or confine smoking to a single well-ventilated room closed off from other sections of the home.

The Bedroom

Lifestyle studies show the average person spends a third of his or her life in the bedroom; the elderly, ill, and very young children spend even more time there. Unfortunately, even those who live alone share their bedrooms with some unwanted guests: microscopic dust mites. Cousins of spiders, these tiny creatures live by the millions in bedding, pillows, furniture, and carpeting and thrive on little more than water and human skin scales, which comprise 70 percent of house dust. In addition, bedrooms contain many of the same chemically treated products that fill the rest of the home: pressed-wood furniture, synthetic fabrics, plastics, and carpeting.

Most older homes provide enough ventilation through leaks around doors and windows to dilute and remove indoor air contaminants from bedrooms. But newer energy-efficient homes can trap contaminants and allow them to build to harmful levels. The best rule of thumb for maintaining a healthful bedroom is to make sure there is enough ventilation before going to sleep, whether natural (open windows) or mechanical (air-handling units). A number of other specific steps can be taken to control pollution emissions at the source.

• To discourage dust mites—whose droppings can produce reactions in asthmatics, allergy sufferers, and healthy people alike—sheets and pillowcases should be washed in hot water once a week. Blankets and pillows should be washed in hot water twice a month. For allergic individuals or asthmatics, allergists advise encasing pillows (non-feather) and mattresses in plastic coverings, and tearing out wall-to-wall carpeting, which encourages mite growth, in favor of washable throw rugs on hardwood floors, hard plastic flooring, or ceramic tile.

• Limit the use of bonded wood products that have not been designed to limit formaldehyde emissions and chemically treated materials in the bedroom. These can include bedding that is dyed or treated to resist staining, mildew, wrinkles, or fire; and pillows stuffed with plastic foams, rubbers, and other synthetics.

• Keep pets, which can carry insects and produce allergenic dander, out of the bedroom.

• A final note: Portable heating units can not only produce combustion gases, as in the case of kerosene, gas, and oil, but they also kick up dust and pose a fire hazard if left on overnight.

The Attached Garage

Homes with attached garages, while convenient, tend to have much higher concentrations of toxic gases and chemical contaminants from a number of sources. According to a 1988 study by Harvard researchers, homes with attached garages had higher levels of benzene, xylenes, carbon tetrachloride, methyl chloroform, paradichlorobenzene, and seven other volatile organic compounds than homes without.[1]

As this study suggests, the garage is the Love Canal of most homes—the place where household chemicals, pesticides, paints, fuels, cleaners, and other toxic products are stored. In addition, the garage typically houses gas-powered lawnmowers, power tools, and, of course, cars that emit benzenes and other VOCs. What's more, a variety of common polluting activities—such as stripping and refinishing furniture and repairing the car—often take place in the garage.

In general, the common strategies of increased ventilation, careful selection of products, and stringent controls of pollution sources are the three best ways to control emissions from attached garages. In addition:

• Take steps to insolate the garage from the rest of the house—weatherstrip around common doorways, for instance—to limit the contaminants that make their way indoors.

• Store chemical cleaners, paints, pesticides, and other products kept in the garage in closed cabinets (vented to the outside, if possible) to keep them away from children and pets. Get rid of older pesticides and chemical products, which may no longer be recommended for use, through designated household hazardous waste collection programs. Buy only as much of a particular product as will be needed to avoid storage problems.

• Open the garage door before starting a car and do not let an idling vehicle sit for long periods of time in an attached garage. Carbon monoxide from an idling car in a closed garage can reach levels as high as 1,500 parts per million within minutes—a concentration that is lethal in under an hour.

Testing Your IAQ

The American Lung Association has produced an indoor air quality (IAQ) checklist to help identify homes that may be at risk for indoor air pollution problems. According to the association, more than ten "yes" answers to the following questions may indicate poor indoor air quality:

- Does the home have any unvented gas appliances?
- Do any household members smoke? (One "yes" for each.)
- Do any furry pets live indoors? (One "yes" for each.)
- Are house plants present?
- Are insecticides or pesticides used indoors?
- How many cars are parked in an attached enclosed garage? (one "yes" for each auto)
- Are any of the following hobbies conducted indoors: woodworking, jewelry making, pottery, or model building?
- Are pressurized aerosol canisters used in the home?
- Is any part of the living area below ground level?
- Is the house insulated with UFFI or asbestos?
- Are heating vents corroded or rusted?
- Do burner flames on gas-heating or cooking appliances appear yellow instead of blue?
- Are there unusual and noticeable odors?
- Is the humidity level unusually high, or is moisture noticeable on windows and other surfaces?
- Does the air seem stale?
- Are any of the following symptoms noticeable among residents: headaches, itchy or watery eyes, nose or throat infection or dryness, dizziness, nausea, colds, sinus problems?
- Is the house temperature unusually warm or cold?
- Is there a noticeable lack of air movement?
- Is dust on furniture noticeable?
- Has the home been weatherized recently?

• Are any family members less than four or more than sixty years old?

• Is anyone normally confined to the house more than twelve hours per day?

• Does anyone suffer from asthma or bronchitis, allergies, heart problems, or hypersensitivity pneumonitis?

CHAPTER 5

A Word About Bigger Problems

Paul Bierman-Lytle has seen the future of the American home. It's in Connecticut.

Bierman-Lytle is a New Age architect who specializes in the construction of what might be called non-toxic homes. The former Rhodes scholar and Yale University graduate heads an innovative development company, The Masters Corporation of New Canaan, that employs state-of-the-art techniques to design and build safe, healthy residences that have fewer indoor air pollution sources and more advanced ventilation systems than the typical new home. The idea, he says, is to build air quality concerns into the design of a new home up front, so that sick building problems are headed off *before* the first board is laid. For Bierman-Lytle, that means equipping homes with high-tech HVAC systems that not only control temperature, humidity, and ventilation levels, but also monitor and remove radon, formalde-hyde, fiber particles, and other contaminants. In addition, it means carefully selecting building materials that don't emit chemical contaminants that can lead to indoor air quality (IAQ) problems down the road, including formaldehyde-free plywood, German-made water-based paints, all-cotton insulation, un-treated and plastic woods, and wallpaper made without chemical dyes, fungicides, or flame retardants.

"We take on projects that we think of as Twenty-first Centu-ry buildings," says Bierman-Lytle, whose company has built some four hundred homes across the United States. "It's really environmental architecture, when you think of it, because what we do not only covers health concerns and indoor air quality, but also energy use, chemical use, and larger environmental issues."

Bierman-Lytle is on the cutting edge of new home design and construction in the 1990s. But he is not alone.

In Minnesota, Honeywell Inc. is spearheading a pilot home-building project that features "smart homes," which use sensors to control humidity levels, temperature, and air changes, as well as advanced heat-recovery systems that conserve 80 percent of the heat typically lost in fresh-air ventilation systems. In Penn-sylvania, the Westinghouse Science and Technology Center of Pittsburgh is developing new environmental monitoring systems designed to both save energy and regulate chemical contam-inants, combustion gases, and other pollution sources indoors. And from Boston to California, the National Association of Homebuilders has helped spawn hundreds of "smart home" proj-ects, featuring domiciles that monitor everything from air quali-ty and energy use to appliance use and security.

"There's a lot of work going on right now on 'smart buil-dings,'" observes Dr. Dale Keairns of the Westinghouse Center. "It goes all the way from new systems that monitor carbon mon-oxide and other contaminants to heating/cooling units that re-cover heat from lighting fixtures to conserve energy."

Although there is a gee-whiz, Jetsons-like quality to many

of the new developments on the smart-home front, the "green" architectural movement is actually grounded in two very simple concepts: Limit individual pollution sources in new home construction and design overall building systems to control and remove those that can't be eliminated up front.

And while many of the high-priced smart home's features may not yet be practical for most Americans, most of the advanced IAQ techniques involved in their design, construction, and maintenance can be incorporated into virtually any home, whether it is old, new, or undergoing renovations. All it takes is a basic understanding of how ventilation units, air pressure, heating/air conditioning systems, building design, pollution sources (indoors and out), and construction materials interact to affect IAQ.

Ventilation

Homes built just two generations ago were far different from those constructed today. Structural shells built in the early part of the century were made from natural materials such as brick, wood, steel, concrete, and stone. And because they tended to be drafty, they didn't require mechanical ventilation systems to remove and dilute indoor contaminants, odors, and water vapor. Today, however, plastic and synthetic building materials are common. And since the 1970s, building shells have become increasingly air-tight to save energy and cut costs.

Consequently, Consumer Product Safety Commission studies indicate the air inside a typical new home today contains traces of more than 150 volatile organic compounds (VOCs)—compared to ten typically found outdoors—from common building materials, consumer products, and household items. What's more, EPA studies have found that the average new house has just a third of the fresh-air ventilation of those built even a generation ago—down from 1.5 air changes per hour (ACH) to 0.5 ACH. In addition, superinsulated homes can have as little as 0.1 ACH per hour. That means the air inside a typical home built a few decades ago was entirely replaced by fresh air every forty minutes, but it can take two hours for the air inside a new home to be exchanged, and up to ten hours for homes tightly insulated with sealant foams, weatherstripping, and vapor barriers.

This decrease in the amount of fresh air infiltrating a home points up the need for increased ventilation in most new and insulated homes. This can be accomplished through either natural or mechanical means.

Natural

Even the most energy-efficient buildings have some level of natural ventilation due to cracks, leaks, and openings around windows, doors, foundation sills, and other areas. Wind and temperature differences between outdoor and indoor air influence the amount of natural ventilation in a home. Winds generally enter (or infiltrate) the windward side of the home and exit (exfiltrate) the other. Circulation can be increased by opening windows on both sides of the house and leaving interior doors open. A natural phenomenon known as the "stack effect"—which causes air to rise from lower levels to upper levels—can also influence natural ventilation. In addition, air pressure differences can determine infiltration and exfiltration rates, particularly in tightly sealed homes. The danger is when too little outdoor air makes its way in to replace indoor air used by combustion appliances (furnaces, stoves, fireplaces). When this occurs, lower air pressure indoors can force outdoor air to be drawn inappropriately back into the home through chimneys and exhaust flues—a phenomenon called "backdrafting," which disperses combustion products throughout the indoor environment. In some cases, radon and soil gases can be also drawn up from building foundations and into the living areas of a home through such forces. Most older homes have enough natural ventilation to ward off such problems. But superinsulated and newer energy-efficient homes may require mechanical ventilation systems to control such factors and maintain proper air pressure levels.

Mechanical Ventilation

Standard in most new office buildings and multi-family housing units, mechanical ventilation systems use fans and ducts to circulate air and encompass everything from exhaust fans for localized areas (such as kitchen range hoods and bathroom fans) to comprehensive centralized heating-cooling systems and whole-house fans. Although mechanical systems provide more control over indoor air quality in homes without the heat and cooling losses associated with natural methods, they can also present a range of problems themselves. Kitchen and bathroom exhaust systems that are not balanced by an adequate fresh-air supply (i.e. an open window) can cause backdrafting. Whole-house fans can dramatically increase negative air pressure indoors, sucking up soil gases and radon and causing unwanted infiltration of outside pollutants. And all-mechanical systems that force cooled and heated air throughout the home must be properly designed, cleaned, and operated to prevent the growth and dispersion of mold, bacteria, and other microbiological agents that can thrive indoors, and to control indoor levels of VOCs from adhesives, ducting, household products, soil gases, and combustion emissions.

The cornerstone of all successful uses of mechanical ventilation is routine maintenance. Some specialists believe that upwards of 50 percent of all IAQ problems could be solved by simply keeping mechanical ventilation systems working up to their design capacity. That means changing filters regularly and routinely inspecting HVAC units to ensure they are delivering proper levels of fresh air to the home. According to the American Society of Heating, Refrigerating, and Air Conditioning

Engineers' 62-1989 residential ventilation standard, that's 15 cubic feet of air per minute (Cfm) per person in livingrooms, 25 to 100 Cfm in kitchens, and 20 to 50 Cfm in bathrooms.

Energy Conservation

It may seem that adequate indoor air quality can only come at the expense of household energy conservation, whether due to the loss of heat and cooled air through open windows or costs for powering up mechanical heating, ventilating, and air conditioning units. In fact, there are a number of ways homeowners can maximize energy use *and* air quality indoors. In new homes, for instance, forced air systems can be installed with recirculating fans, which draw room air through ducts, mix it with outside air, pass it through a filter, and heat or cool it before venting it through the house. Homes can also be equipped with "air-to-air heat exchangers," which transfer the heat from exhausted indoor air to cooler fresh outdoor air through a maze of ducts and vents. While costly, ranging from a few hundred dollars for wall-mounted units to one thousand dollars for whole-house systems, some can conserve up to 80 percent of the energy lost through merely exhausting stale indoor air. Heat pump ventilators, which provide heat in winter and cool air in summer, can also be used to maintain indoor temperatures without compromising air quality.

In addition, there are a range of simple options for conserving household energy that don't require weatherizing techniques that bottle up indoor pollutants. Using compact fluorescent light bulbs instead of standard incandescents for lighting, which generally accounts for about 10 percent of the average home electric bill, can save nearly as much energy as weatherstripping some houses. Replacing an old hot water heater (20 to 25 percent), refrigerator (15 percent), or dishwasher (5 to 10 percent) with newer energy-efficient models manufactured under the tough guidelines set by the National Energy Conservation Act of 1987 can shave hundreds of dollars from the average household electric bill. Many utility companies offer free home energy audits, compact fluorescent bulbs, low-flow showerheads, and rebates on purchases of energy-efficient lighting and appliances. And a handful of nonprofit research groups, such as the Washington-based American Council for an Energy Efficient Economy, also offer low-cost booklets featuring practical advice for cutting home energy use.

Air Filters

In addition to source controls and ventilation methods, air-cleaning devices can be used to improve air quality indoors. Some forced-air heating and cooling units are equipped with filters or air-purifying systems designed to keep indoor air clean. Often, however, furnace filters are simple devices that use fiberglass or fabrics that can become collection centers for biological agents, dust, and other pollution sources if not regularly replaced or cleaned.

Free-standing portable room air filters can be used to supplement such systems to trap particles, dusts, molds, soot, allergens, and even some chemicals and gases, when used properly. They typically rely on filtration or the attraction of charged particles to the air-cleaning device and/or home surfaces. Although the degree of sophistication and efficacy of air filters varies widely among brands, most devices generally aim to remove one of three primary categories of indoor pollutants: particles (small solid or liquid substances suspended in air such as dusts, fibers, mists, mold spores, dander, pollen); radon and its progeny (from rock, soil, groundwater, and building materials); and gases (from combustion appliances, cigarette smoke, infiltrating car exhaust, and volatile organic compounds from paints, building materials, furnishings, cleaners, and other household products).

The three most common types of devices on the market are mechanical filters, electronic air cleaners, and ion generators.[1]

Mechanical Filters

These devices, which can be free-standing units or installed in the ducts of central heating and air conditioning systems, use a fan to force air through a "flat" (or "panel") filter to capture particles in the air. The filters in such systems are usually made of coarse fibers of glass, animal hair, vegetable material, or synthetics coated with oil or other substances that enhance particulate adhesion. Some flat filters use a permanently charged plastic film or fiber that attracts airborne particles. In general, flat filters are good at trapping large particles (fibers, dusts, pollens) but remove only a small percentage of smaller, respirable ones. Mechanical units with "pleated" or "extended surface" high-efficiency particulate air (HEPA) filters, which have a greater surface area and are more densely packed, generally do a better job of capturing respirable-sized particles than flat filters.

Electronic Air Cleaners

These units generate electrical fields to trap charged particles and may also be installed in central heating/cooling system ducts or portable units equipped with fans. These systems use "electronic precipitators," which allow particles to be collected on a plate, or "charged media filters," where particles are collected on fibers in a filter. But some units can also generate ozone, a lung irritant.

Ion Generators

These devices are similar to electronic air cleaners, but they use static electricity to charge particles in indoor air, which makes them stick to walls, floors, tables, and other surfaces in a room that can be cleaned. Some also contain their own collector units. Like electronic units, however, ion generators can also

113

produce ozone.

Some new units on the market combine one or more of these three primary particle removal systems. Others also may contain adsorbents, activated carbon, and other reactive materials that remove gaseous materials, odors, and particles. No uniform federal standards exist to gauge how well portable air cleaners remove pollutants. But the Chicago-based American Association of Home Appliance Manufacturers has developed a consensus standard, approved by the American National Standards Institute, by which some air cleaners have been certified and rated for their removal of tobacco smoke, dust, and pollen. Under the standard, a unit's effectiveness is expressed in terms of its clean air delivery rate (CADR), which measures both the effectiveness of the device (the percentage of pollutant it removes) and the amount of air it can handle.

In addition to this standard, the EPA has made some general observations about the use and effectiveness of air cleaners. A 1990 agency report, for instance, concluded that units containing electrostatic precipitators, negative ion generators, and/or pleated filters tend to be more effective than flat filter units in removing cigarette smoke particles. But the report added that air-cleaning devices should not be considered a cure-all for indoor air pollution problems and may not be very effective in removing radon gas, volatile organic compounds, and some allergens. "The use of air cleaners," the report stated, "should only be considered when the use of other methods to reduce indoor air pollutants (e.g., controlling specific sources of pollutants or increasing the supply of outdoor air) are not successful in reducing pollutants to acceptable levels."[2]

In addition, indoor air specialists have found that several varieties of common houseplants may even do a better job of filtering the indoor air of certain pollutants than mechanical air-cleaning units. Research conducted in the 1980s by scientists at the National Aeronautics and Space Administration (NASA) determined that substantial percentages of volatile organic chemicals commonly found indoors are filtered out by plant photosynthesis. Formaldehyde, benzene, xylene, and tetrachloroethylene were all found to be removed by the leaves and root systems of such common household fixtures as spider plants, golden pothos, philodendron, bamboo palm, chrysanthemum, corn plants, snake plants, English ivy, Gerbera daisies, peace lilies, and Boston ferns.

Construction Considerations

Building a new home poses many daunting decisions for the prospective homeowner. However, as Paul Bierman-Lytle observes, it also affords far greater power in limiting indoor air pollution risks. Experts advise a few key factors be taken into account before and during new residential construction, among them siting considerations and building materials.

Location, Location, Location

Location not only is a determinant of home value, but can also make a difference in health. Indoor air quality can be affected by pollutants in the air, water, and soils near a home. As the residents of Love Canal, New York, and Times Beach, Missouri, can readily attest, proximity to pollution sources—including heavy industry, major highways, groundwater contamination, hazardous waste sites, or contaminated soils—can all influence the air quality in the home.

Consequently, home builders should be as concerned as home buyers when it comes to researching a future home neighborhood. Local health officials, state environmental agencies, the EPA, and environmental advocacy groups can all be valuable sources of information about the past, present, and future uses of areas surrounding particular housing lots and subdivisions. EPA records, including its Toxics Release Inventory and Superfund site list, can provide information on toxic industrial emissions from industries nearby, locations of suspected and confirmed toxic waste sites, and water quality in local waterways. State and local officials can also help map sites of such potential liabilities as contaminated water wells, leaking underground fuel storage tanks, trash incinerators, and fuel-burning power plants.

In addition, several real estate service companies have emerged in recent years to help new homeowners conduct safety checks before building or purchasing a home. The best-known of these—Environmental Risk Information and Imaging Services of Alexandria, Virginia—markets a nationwide property report service that identifies a range of potential contamination sources. For a flat fee, the company conducts a search of state and federal pollution records to provide documentation of such things as dry cleaners, manufacturing facilities, military dumps, underground petroleum storage tanks, and hazardous spill sites located within one mile of a particular property.

Electromagnetic Fields

Although the jury is still out on the health effects of electromagnetic fields (EMFs), several prominent research studies have linked them to leukemia and brain cancer, suggesting they may represent an important indoor hazard. As scientific research continues, many specialists are recommending that homebuilders be prudent in constructing buildings near transmission lines and other EMF sources, at least until more is known about the health risks posed by EMFs.

Simply stated, EMFs are produced by all things powered by electricity—power lines, electrical equipment, household appliances, computer terminals, electric blankets, even motorized bedside clocks. High-frequency EMFs, produced by X-rays and microwaves, have long been known to pose risks in high enough doses, but new concerns have emerged about the affects of chronic exposure to extremely low-frequency EMFs. Neither the scientific community nor the federal government has been able to definitively link extremely low EMFs to health problems, despite tens of millions of dollars in research. But some independent studies have strongly suggested chronic exposure to EMFs may alter cellular function, inhibit the immune system's natural cancer-fighting mechanisms, and promote cancer cell growth.

A recent twenty-five-year study of half a million people by Swedish researchers found children living near high-tension power lines had four times the leukemia risk of others. Similar findings were made in Denver in the 1970s and substantiated in the 1980s, when researchers linked elevated rates of childhood leukemia to electric power lines. And in 1992, the EPA produced a draft report that found evidence, but not proof, that EMFs are linked to leukemia and brain cancer in children. Yet other studies, including a large survey of West Coast utility workers, have refuted such findings, and the nation's major electric utilities have contended that far greater risks are posed by radon, air pollution, and other recognized hazards.

Although experts believe conclusive scientific evidence may be as many as ten years away, with a $65 million nationwide study authorized under the National Environmental Policy Act of 1992, some specialists are advising that at least prudent steps be taken to limit chronic exposures to EMFs as the debate rages. Such strategies are designed to limit average EMF exposures to below 2 "milligausses"—the level at which some studies have shown biological effects. Homes not now constructed near power lines and electrical equipment, for instance, should be built at least a distance of fifty to two hundred feet away from such facilities. In addition, a number of steps can be taken to determine EMF levels indoors and limit chronic exposures indoors:

• Residents living near high-tension power lines can test for EMFs by asking the local utility company, or hiring a private firm, to measure household levels with devices called "gaussmeters." In many communities, electric utilities or health departments provide these services free of charge.

• Because EMF levels can sometimes be higher in certain rooms in a home simply due to the configuration of house wiring, residents can reduce indoor exposures by limiting the amount of time spent in those rooms where transmission lines attach to the residence. Such rooms can be used as studies or guest rooms, for instance, instead of bedrooms.

• After several health reports noted that electric blankets generate high levels of EMFs—and health officials in New York advised consumers not to use them—manufacturers in the late 1980s moved to redesign them to limit EMF exposures to users. Consequently, EMF exposures can be reduced by replacing older models with newer, better-designed ones. Exposures can also be limited by turning them off and unplugging them—thus removing the EMF source—after they have been used to heat up the bed.

• New computer terminals, like televisions and microwave ovens, are also less dangerous than those made in the 1970s because of manufacturing changes. As with all EMF-generating products, however, health experts advise keeping a distance of at least six inches from such items, especially when they are in use for long periods of time.

Building Materials

Just as avoidance is the best strategy for reducing exposures to EMFs and outdoor pollution sources, it is also the best way to limit contaminant exposures from offgassing indoor sources, beginning with the selection of non-toxic and less-toxic building materials during construction and renovation. In many cases, costs for safer products are not significantly higher than those for more conventionally used, chemically treated materials.

Building foundations made from untreated concrete and brick, for instance, are generally safe, stable, and competitively priced, as are building frames of reinforced concrete, brick, stone, and steel. Yet these products do not contain any of the chemicals, preservatives, or formaldehyde glues found in most pressed-wood products and synthetic materials used in building foundations and frames. If improperly sealed during construction, these treated materials can release offgassing chemicals that can persist for months, even years, inside a tightly insulated new home. In many other areas of building construction, renovation, and finishing, environmental architects say careful selection of materials can make a significant difference in indoor air quality. For instance:

• Interior walls and ceilings made of plaster, drywall, brick, and solid hardwood (maple, birch, poplar) emit lower levels of chemicals than plywoods, particleboard (formaldehyde), and vinyl-coated paneling (plastics).

• Doors, windows, frames, and cabinets made of solid wood (soft or hard), metal, and hard plastic laminates (such as Formica) do not emit the same levels of offgassing chemicals as some softer plastics, interior plywood, or particleboard.

• Flooring materials made of solid hardwood, ceramic tile, concrete, and hard vinyl composition are usually safe and more stable than some soft vinyl products, which can release petroleum products.

• Non-petroleum, water-based adhesives, sealants, and mortars are less toxic than urea-based epoxy, plastic resin glues, and solvent cements.

• Water-based and latex paints, varnishes, and polishes contain fewer solvents and petroleum products. A number are also made with natural bases of linseed oil, olive oil, and beeswax. Hardwood paneling and tile also contain fewer chemicals than vinyl and self-stick wall coverings.

• Siding made of metal, solid wood, stucco, and exterior-grade plywood all contain fewer chemical products than chipboard (formaldehyde glues), vinyl siding (vinyl chloride and plasticizers), and asphalt shingles and fiberboard (petroleum).

Bierman-Lytle, echoing other environmental architects, expects the use of less-toxic materials in the home building industry to become more common as concerns about sick building syndrome continue to grow in the years to come. At the moment, however, the notion of pollution prevention remains a futuristic and foreign concept in all but the most progressive building-trades circles.

"The strategy right now by the government, industry, and mainstream architects and builders is to address sick building syndrome by following [building code] guidelines for VOCs, formaldehyde, and other contaminants," Bierman-Lytle adds. "But our approach is to avoid them altogether. That's the philosophy and mission of the company. And I think that's where we're headed."

CHAPTER 6

A Workable Workplace: Building Air Quality

On May 27, 1989, The Associated Press moved a wire story headlined: "No Answers in What Caused Office Workers to Become Ill." The report, which carried a Wichita dateline, sketched a disturbing scenario worthy of any good Stephen King thriller:

> Even after tests on air, food and water, and blood samples from the victims, authorities have yet to find a cause for 62 office workers becoming so ill that they had to be treated at hospitals. "We may never find out," said Jack Brown, acting director of the city and county health department's environmental health division. "We've done everything possible in the analytical viewpoint. We went from roof to basement and didn't find anything wrong."

Brown and others investigating the phenomenon are considering such causes as lingering airborne pesticides and sick building syndrome at the Pioneer Tele-Technologies office where the employees took sick Thursday night. . . . But every test that Brown and other health and emergency service workers performed provided no insight about the cause. "We've tested all of that with very sophisticated equipment and we've found nothing," Brown said Friday.

About 7:30 P.M., a pregnant worker became ill at the telemarketing office in the lower level of a shopping center, and a paramedic was called. Shortly after that, a second pregnant woman became ill and another ambulance was summoned. About 9:45 P.M., the paramedics were called a third time. This time, it was for dozens of people complaining of headaches, dizziness and nausea. Three paramedics and two firefighters also became ill. By 10 P.M., ambulances were taking people to four hospitals and medical centers. But it didn't end there. Workers who went home had delayed reactions. Several were picked up at their homes by paramedics and taken to the hospital.

Business was normal at the shopping center Friday, but the telemarketing company was closed, and Brown said it was unknown when it would reopen. . . .

As sick building syndrome episodes go, the Wichita case was extraordinary in that it demonstrated just how rapidly indoor air problems can spin wildly out of control, even when quick-response efforts are mounted. But the episode itself, which was later tied to a malfunctioning HVAC system, was anything but extraordinary.

In fact, federal environmental officials estimate as many as one-third of the nation's new public and commercial buildings suffer from one or more common sick building syndrome problems, most of them due to inadequate ventilation. Those odds place anywhere from 50 million to 130 million American workers at risk of experiencing the same kinds of problems that plagued the Pioneer TeleTechnologies office building.[1] And the number of indoor air quality (IAQ) complaints among employees in offices and other non-industrial workplaces has been steadily rising since the 1970s.

NIOSH records indicate, for instance, that only six of the twelve hundred workplace inspections conducted by the U.S. Department of Health and Human Services between 1971 and 1978—or 0.5 percent—were for problems related to indoor air quality. Yet from 1978 to 1980, that number leaped to 12 percent and, between 1980 and 1988, rose as high as 20 percent. Since then, the number of IAQ investigations has continued to climb, now accounting for 30 percent of all building inspections conducted by NIOSH.

In addition, IAQ complaints logged by the EPA have soared since the late 1980s in virtually every region of the United States. And in some areas—New England and the Northeast, for instance—there has been more than a doubling of IAQ queries and complaints, from about 50 per month in 1989 to more than 100 (and as many 250) per month in 1993.[2]

Although most of these complaints reflect only minor IAQ cases and concerns, officials at the EPA and NIOSH say they are also seeing a rapid rise in the number of large-scale sick building outbreaks around the country—episodes that have increasingly become front-page news:

• In August 1991, the *Los Angeles Times* reported that the state Department of General Services of California was investigating the air quality inside two new eighteen-story Sacramento office towers after scores of workers employed by the Departments of Health Services and Social Services began complaining of a rash of classic sick building symptoms.

• In February 1992, the *New York Times* carried a story about federal officials investigating a state Department of

Taxation and Finance building in Albany where a plethora of sick building problems sent nineteen workers to the hospital and seventy more to office clinics with respiratory ailments, eye irritations, and dizziness.

• In June 1993, the *Providence Journal-Bulletin* reported that NIOSH officials had advised administrators at Narragansett High School to upgrade its aging ventilation system and rip out water-damaged carpets after students and teachers reported persistent headaches, eye irritations, respiratory symptoms, and fatigue—ailments that miraculously disappeared once they left the school.

News You Can't Use

What all of these reports indicate, of course, is that increasing attention is being paid to the problems of sick building syndrome by government agencies, workers unions, and the media alike. Yet despite the increasing publicity, little practical IAQ information has trickled down to the owners and operators of public and commercial buildings, whose decisions have the most significant impact on IAQ issues.

Federal, state, and local officials across the country say many building managers and safety officers are generally familiar with the phenomenon of sick building syndrome, often because of union grievances, litigation, or news reports. But when it comes to safe building management practices, most haven't the first idea how to incorporate sound IAQ principles into facility maintenance routines or respond properly to sick building syndrome outbreaks.

Fortunately, this state of affairs appears to be changing as the EPA and other federal agencies are expanding IAQ public education efforts, providing more IAQ resources to business owners and facility managers, and placing increasing emphasis on indoor air issues. EPA-sponsored research of office buildings has also steadily increased since the late 1980s, with the agency launching the largest-ever study of IAQ in American office buildings in late 1991. In addition, the EPA has begun to supplement its publication of IAQ bulletins with technical training sessions around the country to provide practical information to building owners and managers, safety officers, and employers. The agency has also produced a self-paced learning module to introduce building personnel to the basics of managing IAQ and responding to sick building crises.

These and other government efforts are rapidly expanding the base of knowledge available to the building managers and other front-line personnel charged with responding to IAQ problems in the workplace. They have also been a primary factor in the growing interest in IAQ matters among many industries, notably office-supply companies, building materials providers, building trades unions, carpet manufacturers and retailers, property developers, and other real estate professionals.

Yet while the EPA has taken the lead in IAQ public education efforts, the agency has adopted a hands-off approach to responding to specific cases of sick building syndrome. In fact, no single EPA division is authorized to conduct inspections of sick buildings or regulate IAQ matters. And currently only a handful of agency employees are charged with responding to IAQ complaints around the country. In the EPA's Region I–Boston office, one of the agency's most aggressive on IAQ matters, only two employees are responsible for coordinating the agency's IAQ efforts in the six New England states that the office monitors; only one of those specialists is on the job full time.

Where the EPA has fallen behind, however, a small band of federal investigators within NIOSH's Health Hazard Evaluation Program has rushed in to fill the gap. The HHE investigators have neither the regulatory might nor the financial resources of the EPA. But what they lack in enforcement power and resources they have made up in energy, commitment, and effectiveness in solving IAQ problems on a practical, day-to-day level. As the HHE's case records indicate, not only are NIOSH investigators having a tangible impact in improving the air quality in the nation's problem buildings, they are also compiling an impressive body of scientific knowledge that is already informing the actions of policy makers, building managers, IAQ specialists, and government regulators.

Into the Gap

From the floor-to-ceiling windows of NIOSH's eleventh-floor offices in downtown Atlanta, Stan Salisbury and Max Kiefer can literally see the problem all around them: On every side, tall buildings tower over the modern city—structures which, with few exceptions, are outfitted with inoperable windows, equipped with huge and unwieldy HVAC systems, and stuffed with thousands of chemically treated materials, combustion sources, biological agents, and other sources of indoor contaminants.

In Atlanta, as in other major metropolitan areas, the proliferation of such buildings has fundamentally changed the face of the American workday in the past half-century. As the U.S. economy has become less dependent on manufacturing and more reliant on information and services, office buildings have become the modern-day equivalent of the factories and sweat shops of the 1920s and 1930s. Since 1950, U.S. Department of Labor statistics show the number of U.S. manufacturing jobs declined from about 50 percent to 32 percent, while employees in service and high-tech sectors climbed to 51 percent. Without question, this shift has accompanied dramatic improvements in working conditions for the average American, but the rise of The Office Worker in the American workforce has not come without a price. In many ways, the trend has created thorny new work-related problems that go well beyond the authority and resources of local, state, and federal agencies to control.

Kiefer and Salisbury, the two-man IAQ investigations team for NIOSH's twelve-state Southeast region, say sick building problems are enormously complex because of the many factors that can lead to episodes. Malfunctioning HVAC systems, inadequate ventilation, polluted air infiltrating a building from outdoors, and a plethora of common indoor pollution sources can all contribute individually and collectively to sick building syndrome. That's not to mention the scientific, political, and economic factors that are driving public policy in the area. "There's still a lot of controversy and speculation over what causes sick building syndrome and how to deal with it," notes Kiefer. "It's not always clear. And usually it's not one single thing but a multitude of problems."

Because of the complexities, federal agencies have had a difficult time even deciding who should be primarily responsible for fielding the complaints of building occupants who feel their offices are making them sick. The EPA, OSHA, the Department of Energy, and the Consumer Product Safety Commission all share some limited responsibility over matters related to IAQ in the nonindustrial workplace. Consequently, the agencies have historically engaged in a game of political "hot potato" on IAQ issues, with the EPA stepping forward only after Congress compelled the agency to research and publicize IAQ risks in its reauthorization of Superfund in 1986.

In the face of all the bureaucratic squabbling and the absence of a clear congressional mandate on IAQ, NIOSH has opted to leave the political debates to the bureaucrats and strike out on its own to solve some of the pressing IAQ problems in office buildings and nonindustrial workplaces around the country. "Basically, we got into it because nobody else was doing anything on indoor air quality," explains Salisbury, a steely-eyed pragmatist who conducted his first IAQ investigation in 1979 when air quality complaints emerged in offices near a Marietta, Georgia, sewage treatment plant. "OSHA won't do it; OSHA can't do it, because they have authority to enforce only occupational standards. And local and state health officials can't do it, although I think if someone at that level could evaluate the overall problems and offer some commonsense suggestions on how to solve some of these things we could probably solve 50 to 80 percent of the problems."

To date, researchers in NIOSH's Health Hazard Evaluation Program have performed over eleven hundred IAQ inspections in public and commercial buildings. The data from these investigations, compiled by workers in NIOSH's four regional offices, have become the basis for much of what is known about IAQ in public and commercial buildings. NIOSH Director Dr. J. Donald Millar, in hearings before a Superfund subcommittee of the U.S. Senate Committee on Environment and Public Works, summarized much of NIOSH's findings in testimony that provided a troubling warning about IAQ problems. "We have not seen a decrease in indoor air problems," he said in perhaps the understatement of his career, "and we are concerned that as the U.S. moves more and more to a service and information economy, with increases in office workers, the problems will increase."

Although Millar said many of the sick building episodes investigated by NIOSH result from several factors, NIOSH's IAQ inspections indicate there are some striking differences in the kinds of problems that plague office buildings and private homes. While many household problems can be addressed by eliminating, modifying, or controlling pollution sources (i.e., radon, lead, asbestos, mold, combustion gases, and chemicals in household products), the majority of problems in large buildings encountered by NIOSH are associated in some way with mechanical heating, ventilating, and air conditioning (HVAC) units.

In fact, 53 percent of the episodes investigated by NIOSH teams have been linked to inadequate HVAC systems, which can foster IAQ problems in a number of important ways. Poorly designed units can allow odors and contaminants to migrate from underground garages or chemical labs, for example, to administrative office areas. Exhausted air can be drawn back into a building through heat recovery systems or as a result of improperly placed exhaust and intake vents. Mold and other microbial agents can build up in HVAC systems and be spread by vent ducts. Adequate outdoor "fresh" air can be restricted—often intentionally in an effort to save energy costs—leading to increased concentrations of carbon dioxide, chemical contaminants, secondhand smoke, and other indoor pollutants. Improper controls on temperature, humidity, and air flow can also cause discomfort and health problems indoors.

As prevalent as ventilation system problems are, however, they are not the only culprits in sick building syndrome outbreaks. Consequently, increased ventilation is not a cure-all for indoor air pollution problems. Among the other key sick building syndrome causes identified by NIOSH investigations teams:

Pollution from Indoor Sources. Accounting for 15 percent of the cases, indoor contaminants can come from a range of sources identified by NIOSH inspectors, including blueprint copiers (ammonia, acetic acid, volatile organics); pesticides; tobacco smoke; combustion gases from cafeterias, laboratories, heating units, and other locations; boiler additives (diethyl ethanolamine); carpet-cleaning agents; foam insulation, particle board, plywood, construction glues, and adhesives (formaldehyde and organic solvents); lined ventilation ducts (fibrous glass); silicone caulking and curing agents (acetic acid); duplicators (methyl alcohol); and signature machines (butyl methacrylate).

Biological Agents. Adding up to at least 5 percent of the cases, microbiological agents are identified as the likely causes in a variety of cases, including outbreaks of "humidifier lung," Legionnaire's disease, and hypersensitivity pneumonitis—all potentially serious respiratory illnesses. Common culprits include molds, bacteria, viruses, protozoa, and other microbial products in ventilation system components; contamination resulting from water damage to carpets or furnishings; and microbes from standing water in vent systems or near outdoor air intakes.

Contamination from Building Materials. Comprising 4 percent of the cases, chemicals in construction materials, furnishings, and other building products are believed to produce a range of reactions including dermatitis and eye, nose, and throat irritation. Common offenders include formaldehyde offgassing from urea-formaldehyde foam insulation, particleboard, plywood, glues, and adhesives; organic solvents from glues and adhesives; contaminants from fibrous glass; and acetic acid from curing agents in silicone caulking.

Pollution from Outside Sources. Accounting for 10 percent of the cases, contaminants infiltrating buildings from outdoor sources include motor vehicle exhaust; boiler gases; products from construction projects (asphalt, solvents, dusts); gasoline fumes infiltrating basements and/or sewage systems from ruptured underground service station tanks; carbon monoxide from basement parking garages drawn into ventilation systems; and previously exhausted air recirculated back inside as a result of improperly locked or located exhaust and intake vents.

Not all cases investigated by NIOSH have resulted in definitive findings. About 13 percent of the cases have stumped NIOSH teams of industrial hygienists, physicians, and behavioral scientists, according to Millar. For the most part, however, NIOSH has been able to build on its ability to identify the culprits in most SBS cases to craft practical strategic methods for investigating and solving SBS problems. "As our experience increased over time, we developed a solution-oriented approach to conduct these [IAQ] evaluations which places a high priority on building ventilation," he said. "This approach has resulted in the best allocation of our resources and has allowed more efficient use of in-field as well as analytical time."

The Big Picture

Echoing Millar, Salisbury describes NIOSH's IAQ work as

an evolving science that grew out of the findings in individual cases. With no clear IAQ inspection guidelines to follow or legal standards to use as guideposts in nonindustrial settings, NIOSH developed its own methods for evaluating sick buildings by using what little authority Congress gave it in 1970 to respond to inspection requests from workers, employers, and building managers in its capacity as OSHA's primary research arm. Interestingly, the sophisticated industrial hygiene techniques typically used to track air quality in the industrial workplace have not proven to be the most effective in getting to the bottom of IAQ problems in most nonindustrial settings. In fact, it is the low-tech methods—interviewing employees and custodians, tracking smoking policies, conducting simple building "walk-throughs," and physically examining HVAC systems—that have produced the most valuable results.

"We started out using traditional industrial hygiene techniques—doing some air sampling and checking for chemicals," Salisbury explains. "But what we found when we did that is usually the levels of what we were sampling were so low, well below OSHA standards, that it wasn't clear that they were causing a problem. So what we've decided is that it's better to look at the whole situation in a building. For example, the first thing we usually do now when we're called in to do an investigation is a walk-through of the building's HVAC system. And we only sample the air as a precaution, usually because that's what people expect us to do and they'll be disappointed if we don't."

Salisbury says air sampling tests are still valuable in turning up certain problems. The presence of volatile organics, ammonia, and ozone can point the finger at some office copiers. Air-sampling devices can also track mold spores and contaminants from renovation projects. And high levels of carbon dioxide, above ASHRAE's one thousand-parts-per-million cutoff, can indicate a building that is too tightly insulated, densely occupied, or inadequately ventilated.

In most cases, however, air sampling is of only limited

value, Salisbury says. Consequently, building managers should be skeptical of commercial indoor air specialists that rely heavily on expensive air sampling techniques in sick building inspections to the exclusion of other investigative techniques. Because there are no uniform standards, guidelines, or regulations for IAQ specialists, many unscrupulous "experts" charge thousands of dollars to test the chemistry of the air but do little to help managers solve SBS problems. "There are known cases where people have come in and measured levels of oxygen and nitrogen in the air," he explains, noting the two gases are primary components of the air around us. "Unfortunately, this country is just ripe with those kinds of folks."

Indeed. According to EPA records, fewer than twelve hundred IAQ specialists—most of them providing asbestos- and radon-abatement services—were providing services nationwide in 1988. By 1992, EPA surveys showed that that figure had jumped to fifteen hundred, with the IAQ services field now constituting a $6 billion industry that is still expanding. Dave Moudarri, of the EPA's Indoor Air Division, explains that the agency has attempted to help building owners and managers keep up with the IAQ industry's expansion by conducting regular introductory training sessions on IAQ fundamentals. These programs are designed to equip managers and owners with at least enough understanding of the issues to make informed decisions in building management and in hiring IAQ specialists for specific problems. The agency also provides lists of scrupulous IAQ specialists—particularly those handling radon and asbestos abatement—to its regional offices to minimize fraud through limited enforcement efforts. But, as with any growth industry, the IAQ field is one that provides vast opportunities for scam artists to capitalize on the relatively low levels of public understanding of IAQ issues.

This is not to say reputable IAQ specialists or HVAC experts are rare. In fact, in many of the cases it investigates, NIOSH refers building managers to HVAC system contractors and IAQ contractors who can solve particular problems turned up by agency inspectors. But in the absence of a good recommendation, Salisbury says the best way to identify unscrupulous firms is to be on the lookout for consultants that hype the risks and liabilities of SBS or claim that the one service they offer will solve all IAQ problems. Some ventilation duct cleaning firms, for instance, advertise their services as central to IAQ. But unless an SBS condition results from blockages, deposits, mold, or other microbiological growth in the ductwork, duct cleaning will have little positive effect on the air quality. As with other building management services, the best defense against IAQ scam artists is to check references and hire companies that have experience in the field and knowledge of local codes and regional climate conditions.

"If you see people who are using scare tactics, then you should be suspicious," he advises. "If there are people out there saying there are a lot of serious health threats that need to be addressed, and that they can solve all your problems, you should be suspicious. There is really no scientific evidence that duct cleaning, for instance, improves air quality. It depends on what the problem is."

On the Lookout for Health Effects

Echoing NIOSH investigators in other regions of the country, Kiefer and Salisbury say the health complaints they receive in sick building syndrome cases are usually varied and nonspecific. There are, however, some relatively common signs and symptoms of building-related problems, as well as some key benchmarks that investigators consider in defining a sick building episode.

A workable definition of SBS, prescribed by NIOSH and the World Health Organization (WHO), holds that a facility is suspect when at least 20 percent of the occupants are complaining of multiple chronic health problems—such as headache, fatigue, dry and sore throat, eye irritation, skin rashes, nausea, shortness of breath, dizziness, and/or cold- and flu-like symptoms—that do not suggest any single medical diagnosis or causative agent. In typical cases, workers complain of a wide variety symptoms that they believe are related in some way to the facility or its operations. This multiplicity of symptoms is a key factor in diagnosing SBS, specialists say, because it is not uncommon for as many as half of the workers in even healthy buildings to report one or two symptoms over a period of several months.

Because many of the symptoms associated with sick buildings mirror other health conditions, office workers often attribute them to job stress, "the bug that's going around," or other factors unrelated to indoor air quality. "We're just beginning to get to the point where people are starting to realize the headache they have at their desk is not because of the report that's due at the end of the day, but because of poor air quality," Kiefer observes. A key factor in diagnosing SBS is whether the symptoms lessen or clear up once workers leave the facility. Investigators also attempt to determine whether the symptoms surfaced in the wake of a specific development—the installation of a new HVAC system, a new renovation project, or a roof leak, for instance.

Two of the most common problems encountered by NIOSH investigators are "stuffy buildings," where workers complain of fatigue, nausea, headaches, and dizziness, usually as a result of inadequate ventilation and/or malfunctioning HVAC systems; and "sick buildings," where workers have respiratory illnesses that can be caused by biological agents—such as bacteria, viruses, and molds in carpets, ceilings, walls, and air-handling systems—and exacerbated by inadequate ventilation.

Other symptoms described by workers can provide clues to specific contaminant sources. A preponderance of skin rashes can suggest asbestos or fibrous glass in the air. Irritation of workers' eyes, noses, or throats can indicate the presence of formaldehyde or cigarette smoke. Cold- and flu-like symptoms can indicate a problem with mold or biological agents.

Salisbury says NIOSH investigators usually interview workers in an effort to single out common complaints. A twelve-question survey developed by NIOSH asks workers about temperature, humidity, odors, ventilation, and other factors affecting IAQ. It also asks employees to list any symptoms or health effects they experience more than twice a week and whether these symptoms clear up during off hours. Personal questions are also posed about smoking habits, allergies, contact lens wear, and job duties. And, in a summary section, workers are asked open-ended questions about the overall office environment.

Salisbury says the questionnaires are marvelous sources of information that help direct investigators' efforts. The questionnaire-interview process is also helpful in turning up worker complaints that have less to do with IAQ than bad employee–employer relationships, which can multiply as rapidly as mold spores in some office settings. While a small percentage of air quality complaints may be due to such situations, Salisbury generally advises building managers and employers to take all IAQ complaints seriously and respond to them quickly and sincerely, even when they suspect the air quality is fine. "It's important that managers have an attitude of general concern about complaints," he says. "You may not ultimately find a problem, but you should at least look."

Other cardinal rules NIOSH recommends for building managers, owners, and safety officers:

• Routinely maintain and check HVAC systems to head off problems before they arise.

• Follow manufacturers' recommendations when using products that contain toxic chemicals, including pesticides, paints, cleaners, office equipment, and machinery.

• Take a proactive approach to anticipate potential SBS problems when renovating or refurbishing office spaces, painting, or using chemicals indoors. That means isolating areas undergoing modifications from other parts of the building and

135

scheduling the work during off hours when fewer workers are around.

Other Obstacles

Although NIOSH has no regulatory authority to enforce its directives, Kiefer and Salisbury say most managers and building owners respond to the recommendations they have made. Often that's because the IAQ cases NIOSH handles—about thirty a year—are worst-case scenarios, referred by OSHA, local officials, or other agencies that have been unable to diagnose the problem. "Many of our requests are last-resort things," notes Kiefer. "After OSHA has been out, and local officials have been unable to find anything, *then* they call us."

Still, a small minority of building managers and employers have thumbed their noses at the investigators, citing the lack of enforceable federal standards on indoor air quality. Recently, for instance, NIOSH inspected a large office building in Birmingham, Alabama, that was experiencing IAQ problems because the structure was built with inoperable windows *and* an HVAC system that had no fresh air intakes at all. As a result, what little fresh outdoor air made its way into office spaces indoors did so totally by accident. To make matters worse, smoking was not restricted at all inside the building.

"We found that carbon dioxide levels by mid-day were typically above 3,000 parts per million," or three times the 1,000 ppm maximum recommended by ASHRAE, he explains. After conducting an IAQ inspection, NIOSH confronted the building's owner and recommended that he modify the system to allow for adequate ventilation. He flatly refused, citing increased costs for energy and the lack of IAQ standards. "The owner said, 'Hey, I'm not required to do this, so I'm not going to do it,' and that was that," Salisbury says.

Unfortunately, the building's owner was standing on solid legal ground, leaving occupants of the building only one recourse: a lawsuit. Because NIOSH is authorized only to conduct research and make recommendations, the agency has no regulatory authority to enforce federal air pollution laws. What's more, there are currently no uniform enforceable federal building ventilation standards in place, only local building codes that require building developers to show that an HVAC system planned for a proposed facility can meet certain ventilation levels as part of the building permit application processes. But unlike provisions for sprinkler systems and smoke detection devices in buildings, these codes generally do not provide for follow-up action after a building is constructed. Consequently, once a facility is up and operating, building managers in nearly all states are not required to show that the HVAC system is actually functioning as designed.

What all that means is that the one federal agency that is currently taking the lead in investigating IAQ problems—NIOSH—has no power or back-up authority on the books to order corrective actions in sick buildings, even when investigators discover serious problems. In Salisbury's view, this fact highlights the need for a more unified federal policy on IAQ—one that designates a leading federal agency to deal with SBS and empowers officials to take action in certain situations.

At a minimum, NIOSH officials have advocated the formal codification of ASHRAE's building ventilation guidelines—last updated in 1989—into state, regional, and federal building codes. Currently, only a handful of states and regions have adopted building code ventilation requirements at least as stringent as those recommended by ASHRAE. Extending those guidelines to all local, state, regional, and federal building codes would not only give NIOSH a solid supporting foundation for its recommendations, but also increase the pressure on building owners, managers, and employers to incorporate sound IAQ principles into their overall management practices. "I think it would be a good first step for the building codes to adopt the ASHRAE guidelines," says Salisbury, noting some regions of the country have already done so.

"If the Southern Building Code [covering some twenty states] referenced ASHRAE 62-89, that would be at least a step in the right direction," he explains. "And even if you don't have the authority to regulate, those references would be a good start because then we'd have a reference point. We could at least say then that the building doesn't meet the code, and that may not sit too well with some of the occupants of the building."

"But there's a lot of controversy over that issue."

Looking to Congress

While NIOSH supports codification of the ASHRAE standard, the federal agency has consistently opposed strict, enforceable regulations on indoor air quality. At least for the time being. "Some agency needs to coordinate IAQ; someone has to take the lead on it," Salisbury says. "But I don't think we have enough data to begin enforcement yet. I think there should be some minimum requirements for HVAC systems and outdoor air ventilation. But when you talk about enforcement, I'm not sure we're anywhere near that yet."

In his remarks to Congress, Millar echoed Salisbury's sentiments, noting that NIOSH will continue to examine and solve SBS problems, but that the agency will not be able to address the mounting numbers of cases around the country alone. In his view, Congress should channel greater resources into federal programs so that standardized models can be developed for investigating and handling IAQ problems—models that will, without question, be largely based on NIOSH's continuing work in the area.

"Increasing office worker awareness and the current shift to office-based, service-type employment will no doubt increase concerns about the indoor air quality in offices and other nonindustrial settings," added Millar. "More research into office building ventilation and its effect on background levels of contaminants will be necessary to provide additional and more appropriate guidelines for the evaluation and control of indoor air quality problems in the future."

Key Factors in Sick Building Syndrome

A howling wind in his face, Andrew MacDougall shouts to be heard over the rush of frigid air that whips around him like an angry indoor tornado. Leaning into the gale, high above the hallowed halls of the Massachusetts Institute of Technology, MacDougall struggles to explain the inner workings of the air-mixing chamber inside MIT Building 36's massive HVAC system which, today, serves as an unlikely classroom to a handful of IAQ students crowded into the wind-whipped facility.

"This is where the fresh air mixes with recirculated air from below," bellows MacDougall, an MIT building maintenance engineer and HVAC specialist. "Indoor air quality problems in buildings are often a result of what happens right here."

HVAC systems, like the one in MIT Building 36, are indeed at the root of most IAQ complaints and sick building problems in modern facilities. According to NIOSH investigators, who have inspected more than eleven hundred "sick" buildings, more indoor air quality problems are due to improperly functioning HVAC units and inadequate ventilation than all other factors combined, including individual contaminant sources, pathways that ferry polluted air to "clean" areas, and building occupants.

Said another way, properly maintaining a building's mechanical heating, ventilating, and air conditioning system is the most important step a facility manager can take to ensure adequate air quality in a large modern buildings and prevent sick building outbreaks.

At the same time, failing to do so greatly increases the likelihood of building-related health problems—including cold- and flu-like symptoms, allergic reactions, and even life-threatening conditions such as Legionnaire's disease and carbon monoxide poisoning. It can also lead to significant economic losses, reduced worker productivity, deterioration of furnishings and equipment, and increased liability.

Priority 1: HVAC Systems

In simple terms, HVAC systems are the mechanical lungs of large buildings. They can be stand-alone units that serve individual rooms or, more likely, centrally controlled systems that provide heat, cool air, and ventilation to all rooms in the facility. By definition, they include all heating, cooling, and ventilating units (boilers, furnaces, cooling towers, air-handling units) as well as the ductwork, filters, exhaust fans, and fresh-air intake units.

In general, the HVAC system in a healthy building not only meets the minimum requirements of local building codes, but also contributes to the comfort, productivity, health, and well-being of its occupants. Odors and contaminants are controlled and dissipated through the introduction of adequate fresh (outdoor) air, air pressure controls, filtering devices, and exhaust fans. Humidity and temperature are properly regulated—never too high or too low. Fresh air circulates freely, so there is no sense of stuffiness or discomfort.

Most HVAC systems achieve these goals by distributing a blend of outdoor air and recirculated indoor air that is "conditioned" (i.e., cooled, heated, filtered, dehumidified, or humidified). They typically come in two varieties: constant-volume and variable-volume systems. As the name suggests, constant-volume units provide a uniform level of fresh outdoor air and control temperature by heating or cooling it, not by changing the volume of air supplied. Variable-volume units tend to control temperature and humidity by altering the amount of heated or cooled air provided to indoor spaces.

Virtually all HVAC systems that deliver "conditioned" air include a handful of common components, all of which can present their own unique IAQ problems if not properly designed and maintained:

Outdoor Air Intakes. Usually located on the return side of an HVAC system's ductwork, intakes are the gateway for fresh outdoor air to enter a building's ventilation system and are therefore a key factor in a building's air quality. Intakes that are improperly located—near building exhaust outlets, outside pollution sources, standing water, or loading docks, for instance—can be chief sources of sick building problems.

Mixing Chamber. After entering through an intake, outdoor air passes through a valve-like damper and is mixed with a predetermined percentage of indoor air in the mixing chamber of a building's air-handling unit before making its way inside. Problems of underventilation can result if the outdoor air damper does not allow enough fresh air into the chamber to mix with the recirculating indoor air.

Air Filters. Once the air leaves the mixing chamber, it is forced through a filter system designed to remove particles and impurities such as dust, bacteria, insects, soot, and pollens. Some, though not most, are also made to remove gases and volatile organics. Designs include "flat" filters (made from fibers of glass, synthetics, or other materials) and "pleated" filters, which have a greater surface area and are more densely packed. Filters are rated by ASHRAE for low efficiency (removing up to 30 percent of dusts), medium efficiency (up to 60 percent), or high efficiency (up to 95 percent). Regardless of their design or rating, all filters require fastidious cleaning and maintenance. Improperly handled filters can not only reduce ventilation rates and increase contaminant levels indoors, but also lead to clogged heating/cooling coils and dirty ductwork.

Heating and Cooling Coils. After the air has been filtered, it is blown across heating or cooling coils that are carefully controlled. IAQ problems from heating/cooling coils typically result from energy conservation efforts that limit the amount of outdoor air allowed into the system during months of extreme cold or heat. In addition, because cooling coils also dehumidify the air, water can collect in drip pans under the cooling coil and allow mold, bacteria, and other biological agents to grow. Managers of buildings equipped with boilers must also be especially vigilant of added risks of combustion gas leaks and particle distribution indoors. Similarly, water chillers and cooling towers, common in many large buildings, can provide ideal breeding grounds for pathogenic organisms, such as the bacterium that causes Legionnaire's disease.

Humidity Controls. Because some buildings, or rooms, require careful regulation of humidity for health reasons (operating rooms in hospitals) or equipment concerns (computer rooms), some HVAC systems include units that add or remove water vapor from the air once it is heated or cooled. ASHRAE recommends keeping relative humidity between 30 percent (below which air becomes dry, increasing respiratory infection risks) and 60 percent (above which mold, bacteria, and other biological agents thrive). Such systems must be carefully controlled, cleaned, maintained, and inspected to head off biological growth.

Supply Fans and Ducts. Once the air is ready for distribution throughout the building, it moves through a supply chamber and into the ductwork, which carries it to the return and exhaust grills ("terminal devices") in the rooms and spaces it is designed to serve. Because a typical office building can contain miles of ducts, building owners and managers should take all possible steps to limit dust, moisture, and other contaminants from entering the ductwork. If ducts become contaminated with molds, bacteria, or other pollution sources, experts advise professional duct cleaning, but only during off-hours when the HVAC system can be temporarily closed down. The EPA and NIOSH also advise the removal of any water-damaged or contaminated ductwork and careful cleaning of any coils or drip pans that may have led to duct contamination.

Return Air Sytems, Exhaust Fans. After the air has made its way into the spaces and rooms of a building, it is collected—along with odors, contaminants, and carbon dioxide—and "expired" via return-air systems and exhaust fans, usually located near the mixing chamber. Many building codes require areas where levels of contaminants are high—bathrooms, kitchens, parking garages, and smoking areas, for instance—to have individual exhaust fans that separately ventilate them to supplement the return-air systems.

Using HVAC Systems to Maintain IAQ

To maintain healthy indoor environments, the most aggressive building codes require HVAC system designs to meet at least the recommendations of ASHRAE, a ninety-year-old trade organization that last updated its building ventilation guidelines in 1989 (ASHRAE Standard 62-1989). In coming up with ventilation standards, engineers generally calculate three essential factors: minimum oxygen requirements per person, removal of expired carbon dioxide and contaminants, and comfort considerations such as temperature, humidity, and odors. At a minimum, oxygen demands alone require about 1 Cfm/person of fresh (outdoor) air ventilation in buildings that have no significant pollution sources. Another 2.5 Cfm/person is required to dilute the levels of carbon dioxide in a building from expired breath. These minimums presume fresh outdoor air is clean and free of pollutants and are calculated to keep carbon dioxide levels below one thousand parts per million, which is about 2.5 times typical outdoor concentrations of carbon dioxide.

But simply accounting for biological demands is not all that is required to ensure healthful air quality indoors. Many other factors determine the acceptability of air quality—some of them easily controlled, some not. High temperature and humidity, for instance, can work in tandem to make occupants feel lethargic and uncomfortable. These twin factors can also lead to potent indoor air quality problems. When indoor air is too hot and humid it promotes the spread of molds, fungi, bacteria, viruses, dust mites, and other biological agents that can cause illnesses, allergic reactions, and asthmatic conditions. What's more, high temperature and humidity can lead to chemical reactions indoors, promoting offgassing of formaldehyde and other VOCs from furniture, carpets, furnishings, and fabrics. On the other hand, indoor air that is too dry can also pose a problem. When relative humidity falls below 10 percent, occupants can experience increased incidence of skin problems, scratchy throats, and respiratory infections. This is because dry air tends to increase the amount of airborne dust and allergens indoors while simultaneously drying out the mucous linings of the throat and sinuses, thus increasing human susceptibility to infection as well as allergic reactions and asthmatic conditions.

In general, excess humidity is best handled by increased ventilation of water-rich areas (restrooms, kitchens) while dryness can be controlled through better insulation and weatherstripping, which reduce the need for excess dry heat. Humidifiers can be used to increase water vapor in the air, but unless they are meticulously cleaned, handled, and maintained they can breed biological agents.

In addition to providing adequate oxygen supplies and regulating humidity, HVAC systems can also be used to control odors and contaminants by manipulating the air pressure inside a building. If less air is supplied to a room or building than is exhausted, that room or building will have a "negative" air pressure, and will tend to pull in air from surrounding rooms and areas—including those containing odors and contaminants. This can be a special problem in mixed-use buildings such as schools, hospitals, and apartment houses where contaminants and odors from labs, kitchens, bathrooms, workshops, smoking lounges, and other areas can be drawn inappropriately into cleaner sections. To achieve "positive" air pressure indoors, HVAC systems should be adjusted to ensure that at least as much fresh outdoor air is supplied to a room as is exhausted. Sometimes this can be achieved by simply opening a window or introducing a new fresh-air source; other times it may require a balancing of a building's HVAC system. "Dirty" rooms can also be isolated from "clean" ones by employing local exhaust systems that trap and control contaminants, often through the deliberate creation of negative air pressure conditions, and vent them to the outdoors. In addition, air-cleaning and filtration devices can be utilized either as part of a centralized HVAC system or as a free-standing unit to control contaminants and odors.

Priority 2: Contaminant Sources

While NIOSH surveys demonstrate most sick building problems can be addressed through improved HVAC operations and increased ventilation, such methods are not a cure-all. In fact, a widely publicized four-building, 1,546-worker study reported in the March 1993 issue of *The New England Journal of Medicine* found that simply increasing ventilation does not always alleviate sick building syndrome conditions.[1] Although the study was criticized by some IAQ experts, who noted the four facilities studied did not fit the classic sick building syndrome definition, few argued with researchers' essential claims that increased ventilation should not be regarded as a "cure" for sick buildings. In fact, a number of specific IAQ problems are best resolved by targeting contaminant sources themselves—whether in tandem with increased ventilation or independently.

The *NEJM* report was not the first to contend that increased ventilation is not a cure-all for sick building syndrome. In 1988, the EPA published an 874-page study of indoor air quality in ten buildings—two schools, four office buildings, three nursing homes, and a hospital—that had reached similar conclusions. Conducted over a five-year period, the report, titled "Indoor Air Quality in Public Buildings," found:

• Volatile organic compounds are present in large quantities inside such buildings, with nearly five hundred different chemicals identified and almost every contaminant present at higher levels indoors than outdoors (some as much as one hundred times greater).

• Key sources of some of the target chemicals are offgassing building materials, including molding, paint, carpeting, adhesives, caulking, telephone cables, flooring, and particleboard.

• Concentrations of fine particle pollutants in the air are up to ten times higher in smoking lounges than in other areas.

The EPA report was among the first to identify building products and equipment as a primary culprit, in many cases at least as important in sick building syndrome as inadequate ventilation. Some industrial hygienists predicted it would fuel efforts to control source emissions as well as improve air exchanges in large buildings. "It would not be surprising," wrote David Swankin, in an analysis of the EPA report for the *American Industrial Hygiene Association Journal*, "to see an 'action' list of suspect materials emerge in some quarters in the very near future."

Although Swankin's predictions have not held true, the EPA and NIOSH have increasingly focused on pollution prevention strategies based on source reduction as a key means of controlling sick building syndrome. The two agencies have also identified a number of key examples of sick building causes that cannot always be controlled by merely increasing ventilation.[2] They include contaminants that originate within the building or HVAC system itself, from outdoor sources nearby, and even from the routine activities of the building's occupants.

Outside Sources

Outdoor air, particularly in urban settings, can be contaminated with everything from natural pollutants—pollens, dust, mold spores, biological agents, radon—to industrial pollutants and smog-causing vehicle emissions. The key to limiting infiltration of outdoor contaminant sources is the proper location of outdoor air intakes. Those placed near parking lots, garages, dumpsters, loading docks, standing water (on rooftops, typically), or even the exhaust fans can all lead to problems indoors. Similarly, buildings constructed on or near soil contaminated by previous uses—old landfills or areas laden with pesticides, toxic and/or petroleum wastes, for example—are sick building syndrome cases waiting to happen.

Inside Sources

Sick building problems also frequently result from contaminants that originate within the building itself, from equipment, supplies, motors, construction materials, and human activities. VOCs and ozone, for instance, are commonly emitted from office equipment. Solvents, toners, and ammonia are frequently released from supplies, cleaning materials, labs, and workshops. Textured carpeting, curtains, floors, textiles, furnishings, and asbestos insulation can produce dust and fibers. Volatile organics, such as formaldehyde, and inorganic chemicals can be released from furnishings, building materials, paints, caulking, and adhesives. Water damage or unsanitary conditions can lead to microbiological growth (mold, bacteria, viruses) and infiltration of sewer gases. And even components of the HVAC system itself can be a problem—with dust and microbiological growth in drip pans, ductwork, coils, and other components; improper use of cleaning materials; and inadequate ventilation of combustion gases. Proper HVAC maintenance, careful selection of building materials and equipment, and isolating contaminant sources can all limit the levels of indoor contaminants from specific indoor sources.

Human Activities

Sometimes personal activities—smoking, cooking, cosmetics, or body odors—can lead to indoor pollution problems. Simple good housekeeping and maintenance—cleaning, use of deodorizers, vacuuming, pesticide applications, painting, remodeling—can also produce high indoor concentrations of chemicals, solvents, dusts, and other contaminants. Restricting occupant activities that produce indoor pollutants to one room, section, or floor of the building and separately ventilating areas of high contaminant concentrations (bathrooms, kitchens, labs, smoking lounges) can significantly improve a building's overall air quality.

Priority 3: Pollution Pathways

As is the case in residential settings, air pressure factors inside large buildings can move airborne contaminants from "dirty" areas to "clean" sections if not properly controlled. Doors, windows, hallways, stairwells, and elevator shafts can all act to override a building's air-distribution system by ferrying pollutants from one end to another. The wind, negative air pressure, and the "stack effect"—the tendency of hot air to rise indoors—can all act to overwhelm a building's ventilation system if pollution pathways are not adequately managed.

The best methods for limiting the impact of pathways in spreading pollution involve controlling negative air pressure through proper ventilation, and sealing off and separately ventilating areas of high contaminant concentration (labs, workshops, parking garages, kitchens). Professional IAQ specialists can also help identify problems by using tracer gases to track airflow between rooms, floors, and sections of a building.

Priority 4: Building Occupants

Almost always, symptoms and complaints among a building's occupants are the very first signs of air quality problems in an office or other large structure. Unfortunately, many building owners, managers, health and safety officers, and employers wait until a flood of health problems and complaints arise before taking occupant grumbles about air quality seriously.

Indoor air studies indicate some sensitive individuals may be more likely to notice air quality problems first and, therefore, should be viewed as sick building syndrome "canaries." These people include occupants with allergies to tobacco smoke, molds, and biological agents; asthmatics and people with other respiratory diseases; those with suppressed immune systems (due to disease or chemotherapy); and those who wear contact lenses, whose eyes are more likely to be irritated by indoor pollutants.

Because these individuals may be especially vulnerable to the effects of indoor contaminants, they are more likely to exhibit symptoms before most other occupants in a building. Although health effects can vary greatly among occupants exposed to particular contaminants and mixtures of pollutants, the most common ailments reported in sick building cases include fatigue, respiratory distress, sinus congestion, headache, rashes, cold- and flu-like symptoms, dizziness, nausea, and eye, nose, and throat irritation.

Of course these symptoms are also signs of many other conditions that are not related to indoor air quality, including job stress and illness. But, in general, when at least 20 percent of a building's occupants report multiple health problems that appear to dissipate when they leave the facility, specialists generally suspect sick building syndrome. A second term, *building-related illness,* generally refers to cases where occupants are suffering from a specific ailment—such as Legionnaire's disease and hypersensitivity pneumonitis—linked to environmental contaminants indoors.

The best way to limit outbreaks of both conditions is to encourage open communication among building occupants and managers. IAQ experts advise immediate action at the first signs of trouble, even when complaints are suspect.

Putting It All Together

In order to deal effectively with all potential building air quality parameters, the EPA and NIOSH experts have come up with some general rules of thumb for building owners and managers:

Encourage active exchanges of information. When a building manager or employer loses the trust of its occupants, even the simplest IAQ problems can escalate to charges of unsafe management practices, labor grievances, and lawsuits. It may be a cliché, but honesty truly is the best policy in building management. That includes promoting an atmosphere of open communications among building staff and occupants.

Activate front-line facility staff. Building maintenance workers are often in the best position to identify real or potential IAQ problems, but if uninformed of IAQ issues, they can be a chief source of IAQ problems. Keeping maintenance staff up to date on IAQ issues is not only crucial in identifying and solving SBS problems, but also critical to preventing them. As with most health matters, the expense and effort required to prevent IAQ problems almost always pales in comparison to the costs and efforts required to solve them once they develop.

Don't sacrifice IAQ for energy savings. Numerous studies show that energy costs associated with operating HVAC systems represent only a small fraction of a building's total energy bill. What's more, energy costs are typically far outdistanced by other expenses, with costs for labor and personnel topping the list. James E. Woods Jr., a professor of building construction at Virginia Polytechnic Institute and State University, has estimated salary and benefits in a typical office building cost $100 to $250 per square foot per year; operations costs add up to about $5 per square foot; and energy costs, $2 per square foot.[3] Unfortunately, many employers have sought to save a few dollars by reducing the flow of outdoor ventilation air only to lose tens of thousands through lost worker productivity, increased absenteeism, and greater liability.

Take all IAQ complaints seriously. Ignoring IAQ complaints is one of the biggest mistakes building owners, managers, and employers can make. Sadly, it is also among the most common. Many IAQ problems are not difficult to solve and can even be handled internally at low cost. Responding immediately and aggressively to IAQ complaints sends the message that the health and well-being of the occupants (employees, renters, etc.) is important and valued. It builds an atmosphere of trust. And it is likely to galvanize a coordinated effort to find solutions. Failing to do so not only sends a negative message, it also opens the door to increased liability.

Get educated. Some building problems are so complicated that even the experts—industrial hygienists, occupational physicians, and mechanical engineers—can be stumped. Understanding the variety of IAQ problems that can arise in specific settings—offices, schools, hospitals, offices, daycare centers, shopping centers, and housing complexes—can go a long way toward preventing them from occurring in the first place. The EPA offers a number of IAQ training programs to acquaint building owners and managers with the particulars of IAQ issues, as do many professional societies. The EPA and NIOSH have also produced several informative guidance documents for facility managers and employers. The best of the bunch is "Building Air Quality: A Guide for Building Owners and Managers," which expands upon many of the issues touched on in this chapter. Reading up on IAQ issues can not only reduce the likelihood of a sick building problem, it can also help increase the odds of recognizing a serious situation and knowing when—and how—to seek professional assistance to resolve it.

CHAPTER 8

A Sick Building Sampler

No two buildings are exactly alike. Consequently, no two sick building cases are precisely the same. Yet a number of common IAQ problems have been identified in a wide variety of buildings, both large and small. What follows is a brief sampling of case records from NIOSH's Health Hazard Evaluation and Technical Assistance Branch and the EPA's Indoor Air Division that demonstrates some of the most frequent causes of sick building syndrome, along with symptoms and some possible solutions.

Problem #1: Not Enough Fresh Outdoor Air

Symptoms: On August 7, 1991, NIOSH investigators were called in to assess the Rhode Island Department of Education Administration Building in Providence after employees complained of fatigue; eye, nose, and throat irritations; headaches;

respiratory ailments; and other health problems. Over a period of several months, sixty-five employees were administered questionnaires and NIOSH officials conducted several walk-through inspections of the one hundred-year-old, three-story facility. Of the fifty-two employees who returned NIOSH questionnaires, 55 percent reported work-related allergic conditions, 59 percent complained of headache and fatigue, 50 percent reported frequent respiratory illnesses, and 46 percent complained of eye irritation. In addition to inspecting HVAC systems and components, inspectors tested the air for asbestos, carbon dioxide, carbon monoxide, lead, and relative humidity.

Problems: Although temperature and relative humidity were deemed acceptable, carbon dioxide concentrations were found to be very close to the 1,000 ppm guideline set by ASHRAE. The reason: the building did not have a mechanical ventilation system for supplying fresh air to the building or exhaust systems in crucial areas, including restrooms. In addition, individual air conditioning units were in disrepair; potential lead, asbestos, and tobacco smoke problems were identified; water-damaged carpeting and ceiling tiles were observed in some areas; and a basement printing press area was found to have oil leaks and contain flammable liquids that might be sources of volatile vapors.

Solutions: To improve overall air quality in the building, NIOSH recommended installing and maintaining a new HVAC system capable of supplying 20 cubic feet per minute (Cfm) per occupant of fresh outdoor air. Inspectors also urged the installation of exhaust fans in restroom areas, restricting smoking to designated areas only, the removal of lead and encapsulation of asbestos materials in the building, replacement of water-damaged carpets and tiles to limit mold growth, and cleaning up leaking oil and liquids in printing rooms.

Problem #2: Contamination from Indoor Sources

Symptoms: In the fall of 1991, NIOSH inspectors

investigated a New York State Department of Taxation and Finance office building in Albany after a series of building-related problems sent dozens of state workers to area hospitals with non-specific symptoms. Dizziness, eye irritation, nausea, skin rashes, respiratory ailments, and series of more bizarre problems, including fainting and vomiting, were all reported by a large percentage of the occupants of the 2,600-worker structure, known as State Building Eight, constructed in 1964.

Problems: Through building inspections, walk-throughs, and air testing, investigators found a host of contaminant sources in the facility that may have accounted for the outbreaks. The aging, overburdened ventilation system was found to be so clogged with filth and paper that some areas of the building received no fresh air at all. Workers were also found to be routinely using the pesticide Dursban indoors to control cockroaches. In addition, a series of other isolated pollution sources were identified, including sulfur dioxide gas from rotting refrigerator coils, hydrocarbon emissions from copy machines, and ventilation system releases of an anti-corrosive chemical (diethylaminoethanol) due to a leaking steam pipe.

Solutions: The state Office of General Services, which manages the building, was urged to take a series of steps to limit toxic agents indoors by controlling them at the source, in addition to upgrading the aged ventilation system. Among them: replacing the hazardous hydrocarbon liquid copy machines; substituting boric acid for the chemical pesticides in controlling insects; and cleaning, updating, and maintaining the HVAC system to prevent future releases of caustic chemicals.

Problem #3: Bioaerosol Contamination

Symptoms: On May 17, 1991, NIOSH received a complaint about the air quality at the University of South Carolina's thirty-two-year-old Administration Building in Tampa. Inspectors reviewed questionnaires administered to 293 employees of the

156

two-story, 75,000-square-foot building by the university's Department of Environmental Health and Safety, as well as seventy-seven workers compensation claims that had been filed in association with health problems allegedly related to IAQ. Investigators found that 70 percent of the workers questioned reported health complaints related to IAQ including headache, eye and throat irritation, and fatigue. In addition, most worker compensation claims originated from the Finance and Accounting Department on the building's first floor.

Problems: In spite of the high numbers of IAQ complaints, inspectors found the building's variable air HVAC system was functioning as designed, delivering adequate levels of fresh outdoor air, adequately controlling temperature, and properly maintaining relative humidity. In addition, air testing did not indicate any significant sources or concentrations of VOCs. What they did find, however, was that bioaerosol monitoring had turned up higher levels of biological agents in the areas where most IAQ complaints had occurred than in other places in the building. In one monitored area, bioaerosol concentrations—notably Aspergillus mold and other fungal species—were found to be even higher than those outdoors. Inspectors identified significantly water-damaged carpet and ceiling tiles in the Finance and Administration office, as well as a musty smell indicating the possible presence of mold, fungus, bacteria, or other microbiological organisms. Additional inspections determined some of the building's air-return vents could be blocked if certain doors were left closed, raising the possibility that bioaerosol contaminants could build up in certain areas of the building.

Solutions: Inspectors recommended the immediate removal of the water-damaged carpet and ceiling tiles. They also urged steps be taken to ensure the building's return-air paths were not blocked when doors were closed in certain areas. That, they suggested, could be accomplished by installing louvers in or over doors that could block return-air pathways.

Problem #4: HVAC Malfunction, Reentrainment of Exhaust Air

Symptoms: In September 1992, NIOSH officials were called in to assess air quality problems in the DuPage County Judicial Office Facility in Wheaton, Illinois, following a long-running series of employee medical problems. Following the lead of a private consultant who had conducted environmental monitoring of the building, NIOSH investigators assessed carbon dioxide levels, relative humidity, and temperature, and checked for the presence of VOCs and pesticides inside the four-story, 350,000-square-foot office building, constructed in the late 1980s. Inspectors also interviewed nearly one hundred of the building's seven hundred employees, including fifty-three of the more than two hundred workers who had reported a variety of symptoms including fatigue, drowsiness, headache, runny nose, sinus congestion, and eye, nose, and throat irritation.

Problems: At the time of its inspection, NIOSH found the building's variable-air HVAC systems were functioning as designed, with carbon dioxide levels below ASHRAE's 1,000 ppm guideline. Relative humidity was determined to be at the high end, but acceptable. Only trace amounts of VOCs, bioaerosols, and other contaminants were found to be present. But reviews of a consultant's report revealed several steps had been taken in April 1992 to modify the building's HVAC system after a humidifier malfunction resulted in the release of water-treatment chemicals indoors, sending twenty employees to the hospital and forcing a building evacuation. After those modifications, inspectors determined, 60 percent less fresh outside air was making its way into the building, and decorative screens on the roof had boxed in both the building's fresh-air intakes and exhaust fans, effectively leading to the reentrainment of exhausted air.

Solutions: NIOSH recommended stepped up efforts to evaluate and maintain the building's HVAC systems to ensure sufficient conditioned outside air was provided. A series of

158

additional advisories were also made: closer oversight of house-keeping, pest control, food preparation, and building mainte-nance operations; proactive identification of projects that may affect IAQ (renovations, redecorating, painting); greater com-munication and education of employees and building contractors about IAQ matters.

Other Common Problems

In addition to the expanded case studies above, NIOSH and the EPA have identified a number of common generic examples of sick building problems and solutions based on more than eleven hundred IAQ building inspections over the last twenty years.[1] Among them:

Problem: Outdoor ventilation rate is too low.

Examples: Occupants report routine odors; complain of drowsiness, headaches, discomfort; peak carbon dioxide levels exceed 1,000 ppm.

Solutions: Increase outdoor air ventilation rates by adjusting or repairing the air intakes, mixing dampers, supplying diffus-ers; increase capacity of heating/cooling coils; install an updated HVAC system that can provide necessary increases.

Problem: Contaminant enters building from outdoors.

Examples: Soil gases (radon, fumes from fuel storage tanks, nearby landfills); air intakes sucking in pollutants from neigh-boring facilities (factories, dumpsters, highways) or other sources (parking lots, standing water on roofs, loading docks, building exhaust).

Solutions: Remove source, if possible (dumpsters, standing water); relocate air intakes; change air pressure to reduce in-filtration of pollutants (install sub-slab depressurization systems to prevent radon and other soil gases from entering the build-ing); use filters to trap incoming pollutants.

Problem: Occupant activities causing IAQ problems.

Examples: Smoking, special activities (cooking, print shops, laboratory work); interference with HVAC system (opening windows, blocking supply diffusers to stop drafts, using space heaters or humidifiers).

Solutions: Eliminate or modify the activity (ban or restrict smoking to separately ventilated areas, place range hoods over lab areas, keep supply diffusers free of blockages); select non-toxic materials (art supplies, printing materials, solvents).

Problem: HVAC system is a source of biological contaminants.

Examples: Molds and bacteria found in drain pans, interior ductwork, or air filters; occupants complain of allergic reactions, cold- and flu-like symptoms.

Solutions: Remove the source (cleaning ducts, pans, outdoor intakes); routinely inspect equipment for signs of corrosion.

Problem: Materials and furnishings produce contaminants.

Examples: Odors from new carpeting, furniture, wall coverings, paints, newly dry-cleaned draperies, cleaners.

Solutions: Increase ventilation to dissipate odors (during installation and for a period of time afterwards); install carpet, conduct renovations and cleanings during off hours; have manufacturers (of carpeting, textiles, furnishings) store them in a dry, clean, well-ventilated area until chemical offgassing has diminished prior to installation; encapsulate sources (seal building materials containing formaldehyde); choose low-emitting or non-toxic products; remove problem items from the building.

CHAPTER 9

Schools, Skating Rinks, and Friendly Skies

In February 1992, sixty-three spectators and eleven players participating in a Wisconsin high school hockey tournament were rushed to hospital emergency rooms with respiratory problems, chest pain, nausea, and other symptoms after combustion gases from a malfunctioning ice-resurfacing machine rose to dangerous levels inside an enclosed skating rink.

In May 1993, administrators at Women and Infants Hospital in Providence, Rhode Island, shut down six operating rooms after nineteen employees suffered cardiac and breathing problems later linked to a chemical contaminant that had made its way into the air-handling unit and was dispersed through sections of the facility.

And in June 1993, school officials in Nashua, New Hampshire, ordered the Searles Elementary School closed after air sampling in four classrooms turned up high levels of methylene chloride—a potential carcinogen used in cleaning products, paint thinners, and aerosol sprays—from an unknown source.

All three cases were noteworthy in that they were sick building outbreaks that received wide local media coverage. They were also notable in that they were textbook illustrations of indoor air quality problems in some of the most sick building–prone indoor environments known to IAQ investigators: multi-use facilities such as schools, skating rinks, and hospitals.

In recent years, the EPA and private indoor air researchers have paid special attention to such facilities because they are uniquely susceptible to air quality problems for several reasons. First, multi-use buildings are places where many different activities—some polluting, some not—all take place under one roof. Some buildings, though confining heavy industrial activities to one wing or section while housing administrative offices in another, do not provide separate HVAC systems. Second, hospitals, schools, and skating rinks tend to be frequented by very sensitive populations: children, seniors, and individuals with prior health conditions.

Despite these realities, most multi-use buildings are not specially designed or built to mitigate the special circumstances they present. Consequently, EPA officials and environmental health specialists have urged administrators of such facilities to pay special attention to the unique set of indoor air quality problems that can plague these buildings, and to take proactive steps to forestall them before they develop.

A Healthy Learning Environment?

Next to the family, the school is perhaps a society's most revered institution. It is a place where children learn not only the three Rs, but also the fundamental principles of good citizenship, social custom, and morality. Unfortunately, the typical elementary, middle, and secondary school is also a source of a great many chemical and biological contaminants. And because children spend up to eight hours a day in school, five days a week, nine months a year, potential exposure risks from these indoor contaminants are not insignificant.

EPA studies have shown, for instance, that nearly all of the nation's 107,000 schools contain some level of asbestos in building materials, insulation, or fireproofing. About a third of those buildings—35,000—are believed to house loose, friable asbestos, posing a potential risk to some 15 million children and 1.4 million staff workers. "Due to a lack of reliable exposure data extracted from epidemiological studies and the absence of an exposure threshold," a 1990 EPA report concluded, "the fact that school children and custodial workers are exposed to any amount of asbestos fibers continues to constitute a concern."[1]

EPA studies conducted in 1992 have also projected that one in five schools may contain unsafe levels of cancer-causing radon gas, based on a nine hundred-school sample. Extrapolating from those projections, the EPA estimated 70,000 classrooms in 15,000 schools have radon levels that exceed the EPA's safe standard of 4 picocuries per liter of air, while 10,000 school classrooms have "excessively high" radon levels. At a congressional hearing in March 1993, U.S. Representative Henry A. Waxman (D-Cal.), chairman of the House Commerce Committee's health and environment subcommittee, said the EPA's findings indicate that for some students, "It may be more dangerous to attend school than it would be to work in a nuclear power plant."

In addition to radon and asbestos, a variety of other indoor contaminants are common in thousands of schools, according to federal studies. Among them:

• *Combustion gases*—including diesel fumes, carbon monoxide, and nitrogen dioxide—from idling buses and vehicles in parking lots near classrooms and fresh air intakes, as well as auto shops, improperly vented boilers, and furnaces.

• *Chemical contaminants*—from pesticides, cleaning

solvents, chemical labs, paints, art supplies, furnishings, building materials, polychlorinated biphenyls (PCBs) in electrical transformers, and other sources.

• *Lead*—in drinking water fountains, especially in buildings constructed before 1986, when the toxic brain-damaging heavy metal was commonly used in plumbing. Levels above twenty parts per billion are deemed unsafe under current EPA guidelines for schools.

• *Biological contaminants*—from leaking roofs, which allow water to dampen insulation, carpets, and woodwork; standing water near fresh air intakes (especially on flat rooftops); and HVAC systems themselves, which can become tainted with molds, mildew, fungal growths, bacteria, viruses, and other microbiologicals. These contaminants pose a health risk to all children, but the nation's 3.7 million children who have asthma are especially prone to health problems as a result of such exposures.

• *Secondhand smoke*—comprised of more than 4,000 chemical compounds—from teachers lounges and other smoking areas, which are often not properly sealed off from other areas and separately ventilated. Secondhand smoke causes 3,000 lung cancer deaths and thousands of lower respiratory tract infections in children each year, the EPA estimates. For these reasons, the agency has urged smoking bans or restrictions in all schools, daycare centers, and other facilities that house children.

In recent years, federal health officials have reported a significant increase in school-related indoor air quality complaints in many areas of the country. A survey of calls to the EPA's Indoor Air Quality Clearinghouse hotline, for instance, found school complaints skyrocketed throughout the Northeast in 1993 after media reports proliferated in the wake of a congressional hearing on air quality problems (primarily radon) in schools. "We've definitely been getting an increased number of calls about problems in schools," says Eugene Benoit, an environmental engineer with the EPA's Region I–Boston office.

MaryBeth Smuts, an EPA toxicologist, added that while the complaints have been most pronounced in New England and Northeastern states, "It's not a state problem, it's not even a regional problem—it's a nationwide problem in the schools."[2]

Schools and educational facilities appear to be uniquely susceptible to IAQ problems for three key reasons. First, they are peopled by children, whose small growing bodies are more vulnerable to the effects of contaminants than adults. Second, they are typically equipped with HVAC systems that require careful (sometimes costly) inspection and maintenance. Finally, the variety and volume of chemicals used in cleaning products, building materials, furnishings, and school supplies have increased over the past few decades, while Americans have attempted to save energy by super-insulating public and private buildings.

Environmental and public health officials believe air quality problems in schools in certain regions have become more common in recent years as shrinking school budgets have limited resources for building maintenance, HVAC system upgrades, and energy use. In Northeastern states, for instance, where health officials typically receive several calls a week about school air quality problems in the late winter and early spring, IAQ specialists have encountered buildings where HVAC systems have been entirely turned off to save energy and money. School buildings have also been found with fresh air intakes located in close proximity to exhaust fans, effectively allowing exhausted or contaminated air to be sucked back into the building. And several cases have been documented where contamination from neighboring outdoor sources—bus exhaust, groundwater plumes contaminated with volatile organics, and leaking landfill gases —have led to school closings.

EPA surveys indicate the Northeast, with its wide swings in temperature and climate, has generally experienced the greatest numbers of complaints about school air quality in the country over the past few years. At the same time, however, the region has also produced some of the most promising solutions. In

Maine, the state chapter of the American Lung Association has proposed a protocol, endorsed by an EPA indoor air task force and adopted informally by other states in the region, for resolving sick building problems in schools. "They're sensible," says the EPA's Smuts, "and they could easily be adopted by other states."

Echoing other IAQ guidelines proposed by the EPA, NIOSH, and private professional organizations for virtually any building, the association's School Indoor Air Quality Plan emphasizes building design, product and building material selection, and ventilation system maintenance—as well as a healthy dose of common sense—as key factors in heading off IAQ problems *before* they arise. Among the specific recommendations for school administrators:

• *School Design.* Mechanical air-handling units should be designed and maintained to meet the consensus indoor air quality guidelines set by the American Society of Heating, Refrigerating, and Air Conditioning Engineers (ASHRAE) for acceptable building ventilation; contaminants from art classes, smoking lounges, and vocational education shops should be contained; and fresh-air intakes should be located away from outdoor pollution sources and exhaust vents.

• *Renovations.* Changes made to building facilities—including structural modifications, minor construction, and painting projects—should be carried out during summer breaks, on weekends, or other off-hours when the school is less populated to minimize impacts. Building ventilation rates should also be increased during and immediately after such projects.

• *Purchasing decisions.* Careful selection and use of nontoxic products indoors—such as water-based and low-emitting art supplies, building materials, carpeting, wall covers, furniture, cleansers, and pest-control products—can limit occupants' exposure to airborne contaminants and prevent problems from developing in the first place.

"The focus right now is on responding to crises," says Norm

Anderson, environmental health director for the American Lung Association of Maine, critiquing conventional IAQ response efforts, "whereas, as a public health agency, our focus is on keeping [them] from occurring."[3]

Skating Rinks

In the fall of 1991, fifty-four people were rushed to the hospital after a "freak accident" at an indoor skating rink near Boston caused a toxic combustion gas buildup during a jammed youth hockey game. In February 1992, a similar accident during a Wisconsin high school hockey tournament sent seventy-four people to hospital emergency rooms with acute respiratory symptoms and related ailments. A month later, a third incident, this time at a Rhode Island skating rink, felled more than a dozen kids participating in a youth hockey tournament.

In all three cases, local health officials initially said unusual, isolated circumstances were to blame—nothing more exotic than malfunctioning ventilation systems or ice-resurfacing equipment. Yet environmental health experts say these three incidents were not unusual at all. In fact, they were classic examples of widespread air quality problems among the nation's one thousand-plus enclosed skating rinks, stemming from the routine emission of combustion gases from the ice-resurfacing machines and other fuel-powered equipment.

According to Dr. John Spengler, an indoor air researcher at the Harvard School of Public Health, air quality problems in indoor rinks are among the most acute air pollution risks facing youngsters who participate in skating activities. This is because the developing lungs of children and teenagers are especially vulnerable to the effects of the toxic combustion gases emitted by ice-resurfacing equipment, particularly nitrogen dioxide. Indoor rinks also provide ideal conditions for dangerous buildups in nitrogen dioxide and carbon monoxide from the fuel-powered resurfacing equipment.

Spengler explains: "What happens is, the ice-cleaning machine goes around the rink every hour or so, or even less during hockey games, and its emissions get trapped inside the enclosed arenas, over that cold ice surface. Some of the levels of carbon monoxide and nitrogen dioxide are fifty times higher than measurements we typically find in houses, and they last for hours, not just brief periods."

Compounding the problem, rink managers in many areas of the country restrict the amount of fresh air ventilation in skating arenas by keeping windows and doors shut tight to prevent outdoor temperatures from melting the ice. As a result, many rinks don't provide enough ventilation to remove and disperse the toxic gases between ice-resurfacing circuits.

A study of air quality in some seventy New England rinks conducted by Spengler and several Harvard colleagues in 1992 indicates the problem may be far more widespread than even the experts had previously imagined. The study, presented at an EPA-sponsored conference of the New England Rink Managers Association in 1991 and published in the *Journal of the Air and Waste Management Association,* found nearly 50 percent of the arenas surveyed had levels of nitrogen dioxide considered unhealthy by the World Health Organization (200 parts per billion for one hour).

"What ends up happening is the ice-cleaning machine goes around the rink and bumps up the levels of pollutants," Spengler explains. "Then, before the ventilation can bring the levels down, it goes around again and bumps the levels up further. So by the end of the day, which for most of these rinks is eighteen hours, the levels by late afternoon or early evening can be very high.

"The other thing is, it's like a canyon in these rinks. You have a cold surface, cold air over that surface and you've enclosed it in glass and boards, sometimes going up to fifteen feet high. I mean, if you tried to design a situation that would trap pollutants in the best way, you'd design this. Then you maximize a person's exposures by making skaters exercise in this environment, and all these things converge."

Concerns about air quality problems in ice rinks actually date back to the 1970s, when reports first emerged on carbon monoxide poisoning among hockey players and spectators. In the wake of those incidents, ice-resurfacing machines were redesigned to include catalytic converters, which control emissions of the combustion gas. But the introduction of the devices has not proven to be a cure-all. First, they must be carefully maintained and checked periodically to ensure proper operation. Second, catalytic converters do nothing to control emissions of nitrogen dioxide.

New concerns about enclosed ice rinks have tended to focus on nitrogen dioxide because it poses a special risk to children that is distinctly different from the hazards posed by carbon monoxide. Carbon monoxide blocks the blood's ability to carry oxygen to vital tissues and organs, causing nausea, vomiting, fatigue, headaches, and, in large doses, asphyxiation. Consequently, seniors and angina sufferers are the most sensitive to the effects of carbon monoxide, even at extremely low levels. Nitrogen dioxide, on the other hand, causes coughing, chest pains, bronchial edema, and eye irritation at low concentrations and increased respiratory infection and decreased lung function at higher levels. As a result, exercising asthmatics—and especially children, who have higher breathing rates than adults—are most at risk from exposure to nitrogen dioxide. "It's clear from the symptoms in our study that nitrogen dioxide is very, very important," Spengler notes, "because with nitrogen dioxide there's a different population at risk. With carbon monoxide it's the [older] guy with angina who's going to have pains and be the most severely impacted. But with nitrogen dioxide it's the young kid, maybe in the pee-wee leagues, who has asthma or is in the preconditions of asthma who's going to be affected and have some severe problems."

To address air quality in indoor skating rinks, the Zamboni

corporation—whose name has become synonymous with ice-resurfacing machines—has begun making new models with better pollution control devices, increased its efforts to inform rink managers of the risks, and is now marketing electric-powered ice-resurfacing machines that emit no pollutants. A Minneapolis, company called Guardian Gas Management Systems has also begun marketing free-standing air-handling units that control emissions of nitrogen dioxide, carbon monoxide, and other contaminants in ice rinks. And two states, Rhode Island and Minnesota, have set specific air quality standards for indoor skating rinks. (Minnesota's standards require immediate corrective actions for levels of carbon monoxide over 30 parts per million and concentrations of nitrogen dioxide over 0.5 parts per million; rink closure and evacuation is required at 125 ppm for CO and 2 ppm for nitrogen dioxide.)

Meanwhile, the EPA has come up with a series of guidelines for rink managers designed to control combustion gases and boost the volume of fresh air ventilation in enclosed rinks. Among the EPA's recommendations for rink managers in all states:

• Conduct regular tune-ups of ice-resurfacing equipment to ensure optimum operations and retrofit any older gas-, propane-, or other fuel-powered vehicles with catalytic converters to control carbon monoxide emissions.

• Ensure heating, ventilating, and air conditioning systems are operating as designed and/or increase the amount of fresh outdoor air introduced into the rink, particularly during resurfacing circuits.

• Warm up fuel-powered vehicles in well-ventilated areas before taking them out on the ice (cold-started catalytic converters are less efficient at removing carbon monoxide than those already running) and limit the number of ice-cleaning circuits.

• Use ice edgers only when rinks are not occupied.

• Consider using an electric-powered ice-resurfacing machine.

In addition, parents, coaches, school officials, and teachers should step into the gap and be aware of the warning signs of carbon monoxide poisoning and nitrogen dioxide exposures. "No rink is immune to these problems," Spengler observes. "And any parent who has a child with chronic wheeze or asthma who is participating in recreational skating, taking skating lessons, or playing hockey should be aware of these problems. Coaches and school officials who ask questions about the air quality in the rinks where their kids play can do a great service by making sure rink managers are very alert to this issue."

Hospitals, Housing Units, and Other Special Environments

In January 1992, NIOSH officials were called in to investigate air quality problems at the Middletown Regional Hospital in Ohio after 25 percent of the employees working in a facility pharmacy complained of dry, itching, and burning eyes, which they attributed to their working conditions. After a standard inspection, officials turned up trace levels of ethylene oxide, a potential human carcinogen and eye irritant used in sterilizing hospital equipment that had apparently leaked into the air-handling system from a neighboring department.

Although the episode was relatively minor and was remedied through upgrades in the hospital's HVAC system, the case demonstrated what IAQ specialists say is one of the most common and daunting problems they face: the sick hospital. "Hospitals are real problem areas and they're often not properly addressed," observes John McCarthy, president of Environmental Health and Engineering, Inc., in Newton, Massachusetts. "Hospitals are pretty unique in that you have a very sensitive population and a lot of potentially toxic agents that are not being properly vented."

Although little data is available on how widespread air quality problems are inside the nation's hospitals, with the EPA yet to conduct a full-scale survey, anecdotal evidence indicates medical facilities may pose the greatest IAQ challenges to specialists and building managers. As McCarthy suggests, this is partly due to the fact that the typical hospital houses a wide variety of indoor pollutants, from infectious viral and bacterial agents, sterilizing agents, therapeutic chemicals, and industrial-strength cleaners and disinfectants to the more run-of-the-mill chemical and biological contaminants that can plague any indoor environment. In addition, hospitals are by definition places that shelter sick and ailing individuals whose systems are already compromised, placing them at greater risk of infection and leaving them more vulnerable to the effects of toxic agents.

Yet while hospitals present unique features in their design and function that make a wide array of IAQ problems possible, most air quality problems in medical facilities can be handled in the same way other sick building syndrome cases are remedied—through source controls, increased ventilation, and communication with occupants. The same is also true for other high-risk facilities identified by the EPA and IAQ specialists—including housing complexes for the elderly, nursing homes, industrial park buildings, waterfront properties, shopping malls, restaurants, and apartment buildings. The trick, the experts say, is recognizing up front that particular facilities may be prone to specific problems—inadequate dispersion of cooking odors and kitchen exhaust from apartment buildings, for instance—and being on the lookout for the first signs of trouble.

Air Pollution in Transit

Until recently, the focus of most public and private indoor air research—and, indeed, this book—has centered squarely on buildings (homes, offices, schools, hospitals, ice rinks, and other facilities), just as outdoor pollution studies in the 1970s and early 1980s tended to target stationary factories and power plants. But since the late 1980s, IAQ researchers have begun to recognize another category of indoor air pollution exposures: mobile sources. Predictably, they have found that many of the same IAQ problems that plague buildings also strike trains, planes, automobiles, and other vehicles.

Harvard researchers have found, for instance, that motorists face among the most dramatic exposures to carbon monoxide, VOCs, and microbiological agents while commuting to and from work. According to a 1991 Harvard study that tracked twenty-five non-smoking Boston commuters, motorists driving cars were exposed to three to five times higher levels of six gasoline-related VOCs—benzene, toluene, formaldehyde, ethyl-benzene, m-/p-xylene, and o-xylene—than participants who walked, biked, or took the subway to work. The study, published in the *Journal of the Air and Waste Management Association*, also found little difference in VOC concentrations among old, new, domestic, or foreign cars. However, the highest VOC concentrations were measured in vehicles using car heaters and traveling urban roadways. Researchers also found VOC levels in cars were typically far higher than in participants' homes and offices. Because Americans typically spend 5 percent to 10 percent of their time in transit, commuting represents a potentially significant exposure route, accounting for 10 to 20 percent of an individual's exposure to the VOCs monitored.

A second 1991 Harvard study reached similar conclusions about in-transit exposures to VOCs, ozone, carbon monoxide, and nitrogen dioxide. Two four-door sedans were used to evaluate concentrations of these contaminants under different driving conditions at the EPA's mobile emissions test facilities at Research Triangle Park, North Carolina. And what the researchers found and reported in *Environmental Science and Technology* was that VOC levels were higher inside the vehicles than outdoors, VOCs and carbon monoxide were highest while traveling urban roadways, and higher levels of ozone and nitrogen dioxide were measured during afternoon driving conditions. In addition, the study found the lowest VOC levels were measured while car air conditioners were running; the highest were detected when vehicle vents were open and fans were on.

As researchers have delved into the indoor pollution risks facing motorists, new concerns are also being raised about the levels of fresh air ventilation provided inside the nation's fleet of commercial aircraft, constructed with energy-conserving features in the 1980s. Unlike planes built before the mid-1980s, which circulated 100 percent fresh air every three minutes or so, many models of new aircraft designed to use less fuel supply only 50 percent fresh air—combined with 50 percent recirculated air—every six or seven minutes. Such ventilation levels are well below the office-ventilation guidelines set by ASHRAE.

The change has produced a flood of complaints of headaches, fatigue, nausea, and other ailments among flight attendants and passengers flying new Boeing 737s, 747s, 757s, and 767s, as well as McDonnell Douglas MD80s and MD11s.[4] If current trends continue, those complaints are almost certain to rise. According to government studies, about half the seats sold on commercial flights today are on aircraft with the new ventilation systems. In 1985, only about 30 percent of the flights were on planes with the newer systems.

One of the problems is that decreased levels of fresh air in crowded airplanes can significantly reduce oxygen levels and boost carbon dioxide in the aircraft cabin. A 1992 NIOSH study, requested by Alaska Airlines flight attendants, found carbon dioxide levels can rise to nearly 5,000 parts per billion—almost five times the ASHRAE standard for buildings —inside the new MD80s. Lowered ventilation rates can also effectively increase the concentrations of combustion fumes and chemicals used to construct, maintain, and clean the aircraft, including volatile organics and pesticides. A 1986 study by the National Research Council of the National Academy of Sciences concluded many such chemicals are toxic and may pose health risks; a 1989 Department of Transportation study noted that carbon monoxide aboard poorly ventilated planes can cause significant respiratory ailments.

In addition to these health risks, numerous studies have found that planes with poor ventilation increase passenger risks of contracting contagious illnesses—from cold, flu, measles, and viral infections to more dangerous diseases such as tuberculosis. In 1979, nearly 75 percent of the passengers on board an Air Alaska flight to Kodiak, Alaska, caught the flu after being grounded for nearly four hours with no ventilation.

Concerns about aircraft ventilation have led to some progress in airline air quality practices in recent years. In 1990, a nationwide ban on smoking on domestic flights—urged in 1986 by the National Academy of Sciences—went into effect. The Federal Aviation Administration has also begun to take some tentative steps toward adopting broader regulations on air quality in aircraft cabins. Measures currently under review would set carbon dioxide standards, based on ASHRAE guidelines, and limit the amount of time a plane is allowed to sit on airport runways with decreased ventilation.

As with other proposed indoor air quality standards, however, the regulations under review by the FAA and other federal agencies are not expected to be implemented any time soon. For the time being, the only steps an air traveler can take to ward off "air sickness" is to open the vents—aptly termed "gaspers"—above the seat and avoid sitting in the back of the plane, where air quality is the worst.

Some Parting Thoughts

What all of the studies and case records in this book suggest is just how complicated some indoor air quality problems are, how varied personal exposure routes can be, and how difficult certain IAQ issues are to solve and address. There is no single cure for sick buildings, no end-of-pipe solution, no single industry to target for action, no overriding regulation that would address all of the nation's IAQ episodes. For these reasons, indoor air pollution represents the most complex and challenging environmental health issue facing the nation's policy makers and political leaders.

Yet experts on all sides of IAQ issues seem to agree that more needs to be done at the federal level to pull together all of the available information on the risks and remedies of indoor air pollution so that a more comprehensive, pro-active approach can be taken to address pressing IAQ problems. The federal response to IAQ issues may—and probably *should*—differ markedly from past government-led approaches to national pollution problems. But there is little doubt that until such an initiative is developed and undertaken, indoor air pollution problems will continue to exact a tremendous toll on the nation's health, well-being, and economic vitality.

In testimony before the U.S. House Commerce Committee's influential health and environment subcommittee in March 1993, Dr. Thomas Godar, a pulmonary physician and past president of the American Lung Association, put his finger on the overarching problem in a way that underscores the sentiments of many in the fields of environmental health and IAQ research. In his remarks, Godar was discussing the specific risks of air quality problems in school buildings. But he made it clear that IAQ risks in schools are merely a small subset of the much larger indoor air pollution problems facing the nation at home, at work, and at play.

Godar's 1993 observations did not differ markedly from those he made during the very first congressional hearings on indoor air pollution legislation back in 1987. Yet even today they resonate with a sense of urgency and clarity. Consequently, they serve as a fitting summary that not only condenses historical IAQ developments and reflects the current state of affairs, but also suggests how the next chapters on domestic IAQ policy should be written.

"Although indoor air pollution is recognized as a public

health problem," Godar asserted, "the government has not fulfilled its promise to protect the public's health from this hazard. To date, no single federal agency has been given the jurisdiction for ensuring healthful air quality in the non-workplace indoor environment [and] an adequate national response to our indoor air pollution problem is long overdue. . . .

"With increased public education and consumer awareness, organizations like the American Lung Association have embarked on successful efforts to address our rapidly expanding environmental needs. While these elements have proven successful, public awareness cannot be effective without a sincere commitment from the federal government to alleviate the health risks posed by indoor air pollution.

"Diverse groups, from health-oriented organizations like the American Lung Association to chemical manufacturers and commercial builders, are urgently seeking federal action to ensure healthier and safer indoor environments in the workplace, at school, and in our homes.

"These groups are looking to Congress and the federal agencies for leadership."

NOTES

Chapter 1

1. *Chemical Exposures: Low Levels and High Stakes.* Nicholas A. Ashford, Claudia S. Miller. Van Nostrand Reinhold, 1991.

Chapter 2

1. *Report to Congress on Indoor Air Quality; Volume II: Assessment and Control of Indoor Air Pollution.* U.S. EPA, Office of Air and Radiation, August 1989.

2. "Introduction to Indoor Air Quality: A Reference Manual." U.S. EPA, Office of Air and Radiation, July 1991.

3. *Comparative Dosimetry of Radon in Mines and Homes.* National Research Council, National Academy of Sciences. National Academy Press, 1991.

4. *The New York Times,* March 21, 1993.

5. *Indoor Air Pollution in Massachusetts.* Commonwealth of Massachusetts Special Legislative Commission on Indoor Air Pollution, April 1989.

6. *The New York Times,* July 7, 1989.

7. *Indoor Air Pollution: A Health Perspective.* Jonathan M. Samet, John D. Spengler. The Johns Hopkins University Press, 1991.

8. *Indoor Air Pollution: A Health Perspective.* Jonathan M. Samet, John D. Spengler. The Johns Hopkins University Press, 1991.

9. R.G. Lewis, Chief, Methods Research Branch, EPA, Research Triangle Park, NC. "Preliminary Results of the EPA House Dust/Infant Pesticides Exposure Study," as reported in *Pesticide & Toxic Chemical News,* May 1, 1991, p. 11.

Chapter 4

1. Cohen, M.A., Ryan, P.B., Yanagisawa, Y., Spengler, J.D., Ozkaynak, H., and Epstein, P.S. "Indoor/Outdoor Measurements of Volatile Organic Compounds in the Kanawha Valley of West Virginia," presented at the 81st Annual Meeting of the APCA, Dallas, Texas, June 19-24, 1988.

Chapter 5

1. "Residential Air-Cleaning Devices: A Summary of Available Information." U.S. EPA, Air and Radiation, February 1990.
2. Ibid.

Chapter 6

1. Woods, J.E., Jr., testimony to Congress, hearing of the U.S. House Committee on Energy and Commerce, Subcommittee on Health and the Environment. April 10, 1991.
2. Indoor Air Trend Report, memorandum based on U.S. EPA Indoor Air Clearinghouse Hotline data, May 5, 1993. U.S. EPA Region I-Boston.

Chapter 7

1. Menzies, R., et. al. "The Effect of Varying Levels of Outdoor Air Supply on the Symptoms of Sick Building Syndrome." *The New England Journal of Medicine,* March 25, 1993. Vol. 328, Number 12.
2. *Building Air Quality: A Guide for Building Owners and Managers.* EPA, NIOSH, December 1991.
3. *HR Magazine,* February 1990.

Chapter 8

1. *Building Air Quality: A Guide for Building Owner's and Facility Managers.* EPA, NIOSH, December 1991.

Chapter 9

1. *Environmental Hazards in Your School: A Resource Handbook.* U.S. EPA. Publication # 2DT-2001. October 1990.
2. *The Boston Herald,* June 6, 1993.
3. *The Boston Herald,* June 6, 1993.
4. *The New York Times,* June 6, 1993.

BIBLIOGRAPHY

Indoor Air Quality

Coffel, Steve, and Karyn Feiden. *Indoor Pollution*. New York: Ballantine, 1990.

Dadd, Debra Lynn. *The Nontoxic Home and Office*. Jeremy P. Tarcher., 1992.

Environmental Hazards: A Guide for Homeowners and Buyers. California Departments of Real Estate and Health Services, 1991.

Greenfield, Ellen. *House Dangerous: Indoor Pollution in Your Home and Office--and What You Can Do About It*. New York: Vintage, 1987.

Indoor Air Pollution in Massachusetts. Final report of the Special Legislative Commission on Indoor Air Quality, 1989.

Mason Hunter, Linda. *The Healthy Home: An Attic-to-Basement Guide to Toxin-Free Living*. New York: Simon & Schuster, 1989.

National Research Council, National Academy of Sciences. *Indoor Pollutants*. National Academy Press, 1981.

Rousseau, David, W.J. Rea, Jean Enwright. *Your Home, Your Health, and Well-Being*. Ten Speed Press, 1988.

Samet, Jonathan M., and John D. Spengler. *Indoor Air Pollution: A Health Perspective*. Baltimore, MD: Johns Hopkins University Press, 1991.

Turiel, Isaac. *Indoor Air Quality and Human Health*. Stanford, CA: Stanford University Press, 1985.

Environmental Protection Agency Documents

Building Air Quality: A Guide for Building Owners and Facility Managers. EPA National Institute for Occupational Safety and Health, 1991.

Directory of State Indoor Air Contacts. EPA Office of Research and Development (updated annually).

Environmental Quality: 22nd Annual Report from the White House Council for Environmental Quality. 1991.

A Guide to Indoor Air Quality. EPA Consumer Product Safety Commission, 1988.

Report to Congress on Indoor Air Quality: Volume II: Assessment and Control of Indoor Air Pollution. EPA Office of Air and Radiation, 1989.

Total Exposure Assessment Methodology (TEAM) Study: Summary and Analysis, Volume 1. EPA, 1987.

Unfinished Business: A Comparative Assessment of Environmental Problems. Overview Report. EPA Office of Policy Planning and Evaluation, 1987.

Additional Resources

Asbetsos: The Safe Buildings Alliance (Suite 1200, Metropolitan Square, 655 15th St. NW, Washington, DC 20005) has produced several informational brochures, including *Asbestos in Buildings: What Owners and Managers Should Know.* EPA and the CPSC have also issued several information guides, including *Asbestos in the Home.*

Energy Conservation: The *Consumer Guide to Home Energy Savings,* including listings of energy efficient appliances and conservation efforts that don't compromise indoor air quality, is available for a small fee from the American Council for an Energy Efficient Economy, 1001 Connecticut Ave. NW, Suite 801, Washington, DC, 20036.

Environmental Tobacco Smoke: EPA has compiled several brief pamphlets and "Indoor Air Facts" sheets that describe the risks and risk-reduction strategies concerning secondhand smoke and other indoor pollutants, which are available through its regional offices or by calling EPA's Indoor Air Information Clearinghouse hotline: 1-800-438-4318. In addition, the agency published its landmark book, *Respiratory Health Effects of Passive Smoking: Lung Cancer and Other Disorders,* in 1992. The Centers for Disease Control and Prevention's Office on Smoking and Health has also established a secondhand smoke hotline: 1-800-CDC-1311. Most state public health departments also provide information on ETS.

Formaldehyde and VOCs: Several publicly available books provide detailed information on risks and reduction strategies for a number of common volatile organic compounds, among them: *The Green Consumer,* by Joel Makower, with John Elkington and Julia Hailes (Penguin Books, 1990); *The Healthy Home,* by Linda Mason Hunter (Pocket Books, 1989); and *The Nontoxic Home & Office,* by Debra Lynn Dadd (Jeremy P. Tarcher Inc., 1992). EPA has also issued several brochures on VOC-laden products, including *Carpet and Indoor Air,* which are available through its indoor air hotline.

Lead: The National Safety Council has established a National Lead Information Center and hotline (1-800-LEAD-FYI), providing state contacts and educational materials. Several brochures and educational booklets are also available from EPA, among them: *Lead Poisoning and Your Children, Testing Your Home for Lead, Questions Parents Ask About Lead Poisoning,* and *Home Repairs and Renovations: What You Should Know About*

Lead-Based Paint. The Alliance to End Childhood Lead Poisoning (600 Pennsylvania Ave., SE, Suite 100, Washington, DC, 20003) has compiled a *Guide to State Lead Screening Laws.*

Multiple Chemical Sensitivity: A number of MCS support groups have formed to provide assistance and physician referrals, among them: the American Academy of Environmental Medicine (P.O. Box 16106, Denver, CO, 80216); the Human Ecology Action League (2421 W. Pratt, Suite 1112 Chicago, IL, 60645, 312-665-6575); and the Canadian Society for Clinical Ecology and Environmental Medicine (479 Roncesvalles Ave., Toronto, Ont M6R 2N4). In addition, several groundbreaking books have been published, including *Chemical Exposures; Low Levels and High Stakes* by Nicholas Ashford and Claudia Miller (Van Nostrand Reinhold, 1991); *An Alternative Approach to Allergies* by Theron Randolph and Ralph Moss (Lippincott & Crowell, Harper & Row, 1980); and two National Research Council Publications: *Biologic Markers in Immunotoxicology* and *Multiple Chemical Sensitivities* (National Academy Press, 1992).

Pesticides: EPA has produced several helpful publications about pesticides, including: *A Citizen's Guide to Pesticides, Termiticides: Consumer Information* and *To Spray or Not to Spray.* The Department of Food and Agriculture's Pesticide Bureau has also produced *A Homeowner's Guide to the Safe and Proper Use of Pesticides.* Many state and local health departments also provide educational information upon request, as do most major pest-control companies.

Radon: EPA has produced a state-by-state list of radon experts and three guidance booklets on the risks and remedies for radon: *Citizens Guide to Radon, Home Buyer's and Seller's Guide to Radon,* and *Radon Reduction Techniques for Detached Houses.* In addition, several books provide valuable practical information on radon: *Radon: Risk and Remedy* by David Brenner (W.H. Freeman and Co., 1989); and *Comparative Dosimetry of Radon in Mines and Homes* (National Research Council, National Academy Press, 1991).